How to access the supplemental web resource

We are pleased to provide access to a web resource that supplements your textbook, *ACSM's Body Composition Assessment.* This resource offers audio-narrated PowerPoint slideshows that provide the reader with an additional resource to understand the content. Each chapter's main points are discussed in a nontechnical way, giving the novice an introduction to even the most complex topics.

Accessing the web resource is easy!
Follow these steps if you purchased a new book:

1. Visit **www.HumanKinetics.com/ACSMsBodyCompositionAssessment**.

2. Click the <u>first edition</u> link next to the book cover.

3. Click the Sign In link on the left or top of the page. If you do not have an account with Human Kinetics, you will be prompted to create one.

4. If the online product you purchased does not appear in the Ancillary Items box on the left of the page, click the Enter Key Code option in that box. Enter the key code that is printed at the right, including all hyphens. Click the Submit button to unlock your online product.

5. After you have entered your key code the first time, you will never have to enter it again to access this product. Once unlocked, a link to your product will permanently appear in the menu on the left. For future visits, all you need to do is sign in to the textbook's website and follow the link that appears in the left menu!

➡ Click the Need Help? button on the textbook's website if you need assistance along the way.

How to access the web resource if you purchased a used book:

You may purchase access to the web resource by visiting the text's website, **www.HumanKinetics.com/ACSMsBodyCompositionAssessment**, or by calling the following:

800-747-4457 . . . . .U.S. customers
800-465-7301 . . . . .Canadian customers
+44 (0) 113 255 5665 . . . . European customers
217-351-5076 . . . . .International customers

For technical support, send an email to:
support@hkusa.com . . . . U.S. and international customers
info@hkcanada.com . . . . Canadian customers
academic@hkeurope.com . . . . European customers

HUMAN KINETICS

01-2019

ACSM's Body Composition Assessment

Timothy G. Lohman, PhD
University of Arizona, Professor Emeritus

Laurie A. Milliken, PhD, FACSM
University of Massachusetts Boston

Editors

AMERICAN COLLEGE
of SPORTS MEDICINE®

HUMAN KINETICS

Library of Congress Cataloging-in-Publication Data

Names: American College of Sports Medicine, author. | Lohman, Timothy G.,
   1940- editor. | Milliken, Laurie A., 1970- editor.
Title: ACSM's body composition assessment / American College of Sports
   Medicine ; Timothy G. Lohman, Laurie A. Milliken, editors.
Other titles: Body composition assessment
Description: Champaign, IL : Human Kinetics, [2020] | Includes
   bibliographical references and index.
Identifiers: LCCN 2018025272 (print) | LCCN 2018026010 (ebook) | ISBN
   9781492586753 (ebook) | ISBN 9781492526391 (print)
Subjects: | MESH: Body Composition | Anthropometry
Classification: LCC QP34.5 (ebook) | LCC QP34.5 (print) | NLM QU 100 | DDC
   612--dc23
LC record available at https://lccn.loc.gov/2018025272

ISBN: 978-1-4925-2639-1 (print)

Care has been taken to confirm the accuracy of the information present and to describe generally accepted practices. However, the authors, editors, and publisher are not responsible for errors or omissions or for any consequences from application of the information in this publication and make no warranty, expressed or implied, with respect to the currency, completeness, or accuracy of the contents of the publication. Application of this information in a particular situation remains the professional responsibility of the practitioner; the clinical treatments described and recommended may not be considered absolute and universal recommendations. The authors, editors, and publisher have exerted every effort to ensure that drug selection and dosage set forth in this text are in accordance with the current recommendations and practice at the time of publication. However, in view of ongoing research, changes in government regulations, and the constant flow of information relating to drug therapy and drug reactions, the reader is urged to check the package insert for each drug for any change in indications and dosage and for added warnings and precautions. This is particularly important when the recommended agent is a new or infrequently employed drug. Some drugs and medical devices presented in this publication have Food and Drug Administration (FDA) clearance for limited use in restricted research settings. It is the responsibility of the health care provider to ascertain the FDA status of each drug or device planned for use in their clinical practice. For more information concerning the American College of Sports Medicine certification and suggested preparatory materials, call (800) 486-5643 or visit the American College of Sports Medicine Web site at www.acsm.org.

The web addresses cited in this text were current as of December 2018, unless otherwise noted.

Senior Acquisitions Editor: Amy N. Tocco
Developmental Editor: Carly S. O'Connor
Managing Editor: Derek Campbell
Copyeditor: Erin Cler
Indexer: Beth Nauman-Montana
Permissions Manager: Dalene Reeder
Graphic Designer: Denise Lowry
Cover Designer: Keri Evans
Cover Design Associate: Susan Rothermel Allen
Photographs (interior): © Human Kinetics, unless otherwise noted
Photo Asset Manager: Laura Fitch
Photo Production Coordinator: Amy M. Rose
Photo Production Manager: Jason Allen
Senior Art Manager: Kelly Hendren
Printer: Sheridan Books
ACSM Publications Committee Chair: Jeffrey Potteiger, PhD, FACSM
ACSM Chief Content Officer: Katie Feltman
ACSM Development Editor: Angie Chastain

We thank the University of Arizona in Tucson, Arizona, for assistance in providing the location for the photo shoot for this book.

Printed in the United States of America   10 9 8 7 6 5 4 3 2 1

The paper in this book is certified under a sustainable forestry program.

Human Kinetics
P.O. Box 5076
Champaign, IL 61825-5076
Website: www.HumanKinetics.com

In the United States, email info@hkusa.com or call 800-747-4457.
In Canada, email info@hkcanada.com.
In the United Kingdom/Europe, email hk@hkeurope.com.

For information about Human Kinetics' coverage in other areas of the world,
please visit our website: **www.HumanKinetics.com**

CONTENTS

PREFACE

Research on the science of body composition assessment has accelerated over the past 50 years, starting with the first Body Composition Symposium held in New York City in 1961. Following the symposium was a landmark two-volume publication of the body composition papers in the fields of human biology, physical anthropology, exercise science, and nutrition and sports medicine (1).

Early pioneers were Siri (sources of variation in body composition estimates), Brozek (scope of the field of body composition), Behnke (densitometry), Wilmore (sports medicine contributions), Forbes and Pearson (whole-body potassium investigations), Roche and Lohman (growth and body composition studies), Lohman and Roche (anthropometric standardization reference manual), Moore (body cell mass), Mazess (bone density methods and development of field), Heymsfield and Ellis (reference methods), and Wang (body composition models). Many other pioneers have built the field of body composition and emphasized the development of methods and their application to health and chronic disease, growth and development, aging, nutrition, exercise science, sports medicine, pediatrics, and other disciplines. The publication of *Human Body Composition* first (2) and second editions (3) and the applied work of Heyward and Stolarczyk (4) stand as capstones in the development of the field.

Building on these efforts, this textbook continues in the tradition of describing advances in the field of body composition assessment, both in children and adults, with applications to the fields of medicine, exercise science, nutrition, growth and development, and geriatrics. Also, the application of body composition assessment to health and chronic disease with estimates of total body fat, fat distribution, muscle mass, and bone density is an especially essential aspect of the field and this text.

In this text you will find

- descriptions of body composition methods for use in both laboratory and field settings,
- carefully described protocols for the standardization of each method, and
- advantages and limitations for each method following a standardized protocol.

The text is directed toward both the researcher and the clinician to provide a comprehensive approach to the assessment of body composition methodology. Critical to this is the understanding of the sources of error inherent in each measurement technique and to whom these techniques can be applied with accuracy. Given the increasing availability of body composition devices marketed directly to consumers without the context of measurement error and variations within populations, clinicians and researchers play an important role as experts because they aid clients in interpreting body composition results.

Accompanying the text are audio-narrated PowerPoint slideshows (available in the web resource at www.HumanKinetics.com/ACSMsBodyCompositionAssessment) that are intended to provide the reader with an additional resource to understand the content. Each chapter's main points are discussed in a nontechnical way, giving the novice an introduction to even the most complex topics. The text and accompanying slideshows provide a solid foundation for further study in the field.

The authors have had extensive experience in the field of body composition assessment and expect the book to be useful to students of body composition assessment in many fields of study.

ACKNOWLEDGMENTS

We would like to thank the American College of Sports Medicine for the opportunity to write this book. We would also like to acknowledge those researchers in the field of body composition who have created such a rich knowledge base in this area for us to review and explain to the readers. Also, we want to recognize the efforts of the authors of each chapter for their excellent work. This book would not have been written without the organization and talents of Michele Graves, who worked tirelessly on edits and multiple revisions. Finally, we would like to thank the editors at Human Kinetics for their speedy and accurate work to make a great final product.

1

Introduction to Body Composition and Assessment

Timothy G. Lohman, PhD

Laurie A. Milliken, PhD, FACSM

Luis B. Sardinha, PhD

LEARNING OBJECTIVES

After completing this chapter, you will be able to do the following:

- Characterize body composition methods into four levels of accuracy
- Differentiate precision, objectivity, accuracy, and reliability for measurement of body composition
- Describe the concept of technical error of measurement and how it is calculated from a set of observations
- Describe validation and cross-validation studies
- Define the essential terms and concepts for measurement of body composition

There are many methods for assessing body composition, and each method has its advantages and limitations. To better deal with the wide variety of methods, we have organized our approach by categorizing all methods by level of accuracy in estimating body fatness (table 1.1). The level 1 category encompasses the reference methods that are the most accurate. These methods include the multicomponent models, such as measuring body density, body water, and bone mineral, which are covered in chapter 2, where various reference methods are described along with their theoretical and clinical significance.

Table 1.1 Body Composition Methods by Level of Percent Fat Accuracy

Level 1 reference methods (1%-2%)	Level 2 laboratory methods (2%-3%)	Level 3 field methods (3%-4%)	Level 4 field methods (5%-6%)
Magnetic resonance imaging (MRI)	Water displacement densitometry or air displacement plethysmography	Skinfolds	Body mass index (BMI)
Computed tomography (CT)	Dual-energy X-ray absorptiometry (DXA)	Bioelectrical impedance analysis (BIA)	Body size indexes
Multicomponent models	Ultrasound	Circumferences	
	Body water		
	Total body potassium counting		

The second level of body composition methods is the laboratory methods, and they are described in chapter 3. In many instances, these methods are not accurate enough to be classified as reference methods. They are, however, very useful and are often used as criterion methods to evaluate a new method. For example, in a recent survey conducted by the Ad Hoc Working Group of the International Olympic Committee (IOC) Medical Commission dealing with body composition, health, and performance, dual-energy X-ray absorptiometry (DXA), as a laboratory method, was the second most common body composition technique used in athletic populations (1).

The third and fourth levels are the field methods. The third level is more accurate than the fourth level in estimating body fatness. Anthropometry and BIA are among the level 3 methods that are covered in chapter 4. Meyer et al. (1) found that skinfolds and BIA were the first and third most common body composition techniques, respectively, used by an international sample of professionals working with athletic populations. Further description of the methods by level of accuracy is covered by Ackland et al. (2).

Essential to the understanding of body composition assessment are the explanation of concepts covered in this chapter; the assessment of body composition measurement errors, as described in chapter 5; and the application of valid body composition equations for a given population, as described in chapter 7. These key chapters are designed to prevent the misuse of body composition methods in many field situations. Also, an important application of body composition assessment is the estimation of minimum weight (chapter 6). In both the athletic population, where a minimum amount of fat is often the goal of a high-performing athlete, and the eating disorder population, where low body fat puts the individual at greater health risk, accurate estimates of minimum weight are essential and not always easily obtained.

The next part of this chapter is designed to give a more complete understanding of the underpinnings of body composition methods to aid in minimizing errors in their use.

Errors in Body Composition Measurement and Assessment

It is important to keep in mind that all body composition methods are not error free. In a study carried out by several researchers many years ago (3), four experienced investigators measured a small sample of female athletes with four different calipers at five skinfold sites, as defined by Jackson and Pollock (4). The four experienced investigators had not trained together and followed their own interpretation of the skinfold instructions from the published photos and the protocol description from the Jackson and Pollock article (4). The authors found that the mean fat content ranged from 14% to 28%, depending on the caliper, investigator, and equation selected (five different equations were selected representing different populations). In all, there were 80 combinations of investigator, caliper, and equations, and this variation may represent the situation today in the field when there is no standardization of the methods in use. A similar study could be done using BIA with the many units and formulae in use, and it would produce a wide variation in results. Much of the variation in field methods can be reduced if a valid prediction equation with a standardized protocol is used and training procedures are in place (chapters 6 and 7).

PRACTICAL INSIGHTS

Given that the mean percentage of body fat for a sample of participants ranged from 14% to 28% just by virtue of selecting different calipers, investigators, or equations, it is best to perform standardized measurements using consistent protocols to eliminate this source of variation. This variation also exists if you vary methods when performing repeat measurements. It is best practice to keep your methodologies as consistent as possible in addition to being properly trained in carrying out those methods. This will limit your technical measurement errors and increase your ability to detect actual body fat changes over time. When possible, perform repeat measurements using the same instrument and the same equation, at the same time of day, and by the same trained technician for the lowest error.

A key study illustrating this finding is summarized in the monograph *Advances in Body Composition Assessment* (5) in which a standard protocol and training procedure were followed among six research laboratories with each laboratory measuring 50 young adults using both skinfolds and BIA; these measurements were then compared to measurements obtained from underwater weighing, which was the criterion method used during the study. For both skinfold and BIA methods, the standard error of estimate (SEE) was 3.5% fat, and similar results were found among the six laboratories for both methods. The results contrast with many studies where different protocols were followed by different investigators.

As illustrated in the preceding text, it is important to minimize errors when measuring and assessing body composition. Ultimately, there are two main sources of error that need to be considered: the technical error associated with the measurer performing the measurement and the biological error associated with the method that is chosen. Technical errors relate to the concepts of precision, reliability, and objectivity, whereas the biological error inherent in the method chosen relates to accuracy and validity. The concepts of precision, objectivity, and reliability are fundamentally important in ensuring that we diminish the error associated with the method that we choose. Education and training are essential in minimizing technical errors.

PRACTICAL INSIGHTS

The standard error of estimate (SEE) is a common error reported in studies that compare a new body composition technique to an established technique. The smaller the SEE, the better the new technique estimates the established or reference method. An acceptable SEE for percent body fat is 3.5% or smaller when compared to an established technique. If the SEE is larger than 3.5%, this means that the new technique has an error rate that is too high to be acceptable. This is one value to look at when evaluating whether a method is an accurate method to use to measure percent body fat. Sometimes a field technique is all that is available to a practitioner; some field techniques can have higher errors. This needs to be considered when interpreting the final values obtained. If you measure someone's percent body fat using a technique that has an SEE of 5%, their actual value could vary 5% higher or lower than the value you measured. For example, if you measured someone at 23% fat using a method with a 5% SEE, their actual value could be as high as 28% or as low as 18% in two-thirds of the cases. In one-third of the cases, the error would be even larger. The smaller the SEE, the better your estimate of someone's percent body fat.

Technical Errors

To understand how to minimize technical errors, it is important to understand the concepts of precision, objectivity, and reliability. Precision is the degree to which the same measurement under the same conditions (and on the same person) produces the same results (also called reproducibility, or repeatability). A high precision means there is low variability in successive measures, and it is usually represented in the same units of the variable that is being assessed. A high precision doesn't necessarily mean that a method is accurate; for example, a method can yield consistently similar results, but they can be far from the actual value.

Although precision reflects the agreement of repeated body composition measurements of one technique on the same participant, reliability of a technique affords an assessment of measurement error as an amount of variance between participants. The use of reliability instead of precision values has the advantage that the measure has no units; it is therefore possible to compare the precision of variables with different units. The most commonly used measure of reliability is the intraclass correlation coefficient (ICC), with values ranging from 0 (not reliable method) to 1 (perfectly reliable method).

Objectivity is the level of agreement between technicians (intermeasurer reliability). It allows the assessment of the variation of measurements performed by different evaluators in the same group of people. An example that may lead to a lower objectivity when using anthropometry is the variation among technicians when marking anatomical points and the inconsistency of the measurement technique. Intensive training with all evaluators prior to a study is therefore necessary when using multiple evaluators to ensure a high objectivity of the data that are being collected.

Measures of precision, objectivity, and reliability are often dependent on the population observed because the size of the measure may affect the size of the error. Therefore, when assessing the error of a technique, the evaluator should use a population with a morphology similar to that of the population that is going to be assessed in the investigation. For example, if the investigator is going to conduct a study with persons of moderate to high body fatness, the precision, objectivity, or reliability of the equipment should not be assessed in athletes, who normally present an increased muscle mass development with low levels of fat mass (FM).

Although errors associated with body composition techniques are unavoidable, we can minimize the errors by choosing equipment with higher precision. Additionally, we can diminish other sources of error by standardizing procedures, performing regular calibration of the equipment, and ensuring that all technical staff members receive proper training before collecting any data. Finally, we can measure the repeatability of our measures by calculating the error associated with these repeated measurements.

The most commonly used measure of precision and objectivity is the technical error of measurement (TEM), described in equation 1.1. The TEM can be used to assess precision, when the same evaluator produces repeated measures, or objectivity, when different technicians assess the same person with the same technique under similar conditions. This measurement of quality control is crucial for the evaluator to be able to improve the performance of the assessment procedures through a training process. Additionally, it is necessary for TEM to be reported on investigations with repeated measures to analyze whether the observed differences are superior to TEM. Otherwise, the observed differences may not be attributed to exposure/intervention but to the TEM. In the case of precision, where you have two measurements made by one observer on each of several subjects (n), the TEM is

$$\text{TEM} = \sqrt{\frac{\sum\left(D^2\right)}{2n}} \tag{1.1}$$

where D is the difference between the two scores for each person.

In the case of objectivity, technical error is calculated from two measurements, one from each of two observers on each of several subjects (*n*). *D* in equation 1.1 would be the difference between scores measured by the two observers on the same person. A smaller value for the TEM means there is less error in the repeated measurements made by the same person (precision) or made by different people (objectivity).

In addition to the previously mentioned ICC to quantify reliability, the standard error of measurement (SEM) can also be determined. The SEM is a measure of reliability that provides an indication of the amount of variation (dispersion) of the measurement errors (difference between repeated measures using the same method). Specifically, two measurements are made on a group of people by the same technician using the same method. The difference between the measurements for each person is calculated. The SEM is computed by multiplying the standard deviation (SD) of the differences between two repeated measures by the square root of 1 minus the reliability (*r*) (e.g., the ICC; equation 1.2).

$$SEM = SD \times \sqrt{1 - r} \qquad\qquad (1.2)$$

For example, if five people were measured twice by a single technician and the differences between these repeated measurements were −2.4, −0.3, +3.7, +0.9, and +1.5, respectively, and the ICC was *r* = 0.90, the SEM would be calculated as follows. The SD of these differences between measures is 2.25. Therefore, substituting these values into equation 1.2 gives SEM = 2.25 × 0.32 = 0.71.

Measures of precision, objectivity, and reliability are recommended for every body composition assessment study to describe the measurement, equipment, and technicians from which they were calculated. For example, the TEM for laboratory X for DXA model X for evaluator X for FM cannot be applied to laboratory Y, or a model Y, or an evaluator Y for FM, even when measured under the same conditions.

Systematic error is error that is not random but introduced through variations in measurement technique among measurers or faulty calibration of a measurement device. The error is additive, and the systematic error therefore arises from several biases in measurements that when combined lead to a situation where the mean will differ from the actual value. There are several sources of systematic error; the most common examples are calibration of the equipment, the technique of the evaluators, environmental conditions that may interfere with the measure, or when the participant does not meet the prerequisites for a certain body composition assessment. It is very difficult to eliminate systematic errors after the data are already collected because it is difficult to know the size and direction of the errors. We can be dealing with a constant error; for example, when using anthropometry, our caliper always measures +1 mm (+0.04 in.), which is very easy to detect and solve. But we can be dealing with other errors that are more difficult to detect and solve. A common example of systematic error is the time of the day when we assess a person's height. In the morning, the height is slightly higher than at the end of the day. To avoid the time-of-day effect, we need to measure all participants at the same time of the day, preferably in the morning.

If we are aware of the main sources of systematic error, we can easily increase the accuracy of our data. For example, systematic error can be reduced by always calibrating the equipment; giving proper training to the evaluators and standardizing protocols; or even considering other sources, such as environmental considerations or the prerequisites that are fundamental to a given protocol.

Accuracy is the degree to which a given method can estimate the actual value. In general, field methods of body composition (such as skinfold thicknesses or BIA) are 3% to 4% accurate in estimating percent body fat compared to a reference method. The validity of a method is its ability to measure what you intend to measure. Skinfolds measure the subcutaneous thickness of body fat fairly well (accurately); however, they are less valid as a measure of total body fat. There are some assumptions for the use of skinfolds to estimate body fat: a consistent fat patterning, a fixed subcutaneous-to-internal fat relationship, a constant skinfold compressibility, a constant skin-to-adipose fraction, a lipid fraction, and water content of adipose tissue. However, there are interindividual differences in these assumptions that make the use of skinfold measurement to estimate body fat less valid.

The root mean squared deviation (RMSD), also called the pure error, is a measure of the accuracy of the predicted value when comparing to a reference value. To test accuracy, a sample of people is measured twice, once with an established reference method and again with a new method of body composition whose accuracy is being determined. If the new method is accurate, the values obtained will be close to those from the reference method. The formula for calculating the RMSD (equation 1.3) is

$$\text{RMSD} = \sqrt{\frac{\sum \left(\text{measured} - \text{reference}\right)^2}{n}} \qquad \textbf{(1.3)}$$

In table 1.2, the sum of the squared differences is 14.36 (31.65), rounded to 14.4 (32), and we have 10 participants (equation 1.4).

$$\text{RMSD} = \sqrt{\frac{14.4}{10}} = \sqrt{1.44} \pm 1.2 \text{ kg} \qquad \textbf{(1.4)}$$

Therefore, TBW measured by BIA is accurate to ±1.2 kg (±2.6 lb) compared to the reference method (by dilution).

The RMSD can also be used to quantify the difference between an alternative and a reference method by calculating the square root of the average of the square of the residuals (difference between alternative and reference method). By considering the squared root of the errors, it allows for the use of the RMSD, which may be a better estimator of the accuracy of a method for biased samples because it includes both random and systematic errors. The RMSD is an indicator of how well the values assessed by an alternative procedure compare to the actual values assessed by the reference method. A smaller RMSD corresponds to a better accuracy of the alternative procedure. A larger RMSD may arise from differences in the cross-validation sample in terms of age, ethnicity, and sex, for example. It is important to consider that both the SEM (precision) and the RMSD (accuracy) are scale dependent and therefore it is not possible to compare the precision or accuracy of methods that are not assessing the same variable.

Another way to assess the agreement between a reference method and a new method is to quantify the relationship between the two measures. Often, the statistic

Table 1.2 Comparing a Reference Method With an Indirect Method of Assessing Total Body Water (TBW)

Participant	TBW in kg (lb) from BIA (measured)	TBW in kg (lb) from dilution technique (reference)	Difference	(Difference)^2
1	56.1 (123.7)	54.2 (119.5)	1.9 (4.2)	3.61 (7.96)
2	69.2 (152.6)	70.0 (154.3)	−0.8 (−1.8)	0.64 (1.41)
3	47.3 (104.3)	48.1 (106.0)	−0.8 (−1.8)	0.64 (1.41)
4	59.1 (130.3)	59.5 (131.2)	−0.4 (−0.9)	0.16 (0.35)
5	71.8 (158.3)	71.4 (157.4)	0.4 (0.9)	0.16 (0.35)
6	56.5 (124.6)	57.1 (125.9)	−0.6 (−1.3)	0.36 (0.79)
7	60.3 (132.9)	60.0 (132.3)	0.3 (0.7)	0.09 (0.20)
8	53.7 (118.4)	53.2 (117.3)	0.5 (1.1)	0.25 (0.55)
9	51.5 (113.5)	48.6 (107.1)	2.9 (6.4)	8.41 (18.54)
10	59.3 (130.7)	59.1 (130.3)	0.2 (0.4)	0.04 (0.09)
Sum	—	—	—	**14.36 (31.65)**

used in measuring the degree of relationship between two methods of estimating body composition is the correlation coefficient. The correlation varies from −1.0 to 0 to +1.0, depending on the closeness of the relationship. In the body composition field, a low correlation is between 0.1 and 0.3. A high correlation is above 0.8. If we square the correlation, we get an estimate of the variation accounted for in one method by another (coefficient of determination; see table 1.3). The more variation that can be accounted for, the more related are the scores.

The correlation can vary with the variability of the sample under study and is unbiased when a random sample of the population is used. Otherwise, if the sample is less heterogeneous than the population, the population correlation can be underestimated, as is often the case in our volunteer sample characteristics of most research studies. A better estimate of the relationship between two variables is the SEE, which is the SD around a regression line or line of best fit between two variables, which is less dependent on sample variability.

Table 1.3 Squaring the Correlation

Correlation	$r^2 \times 100$ (coefficient of determination)
$r = 0.3$	9%
$r = 0.5$	25%
$r = 0.7$	49%
$r = 0.9$	81%

PRACTICAL INSIGHTS

The coefficient of determination is the $r^2 \times 100$, where the r comes from the relationship between the values of percent fat from a new technique and an established technique. In studies where a new technique is tested, people are measured using both the new technique and an established technique. If the new technique is accurate, the percent fat from the new technique should be similar to that measured by the established technique. Using a statistical technique called regression, you get the r, which shows how related these values are. They should be strongly related if they are accurately measuring the same thing (percent body fat). For a new technique to be considered accurate, the coefficient of determination should be greater than 80%. Earlier in the chapter, we saw that the SEE should also be low. In addition, while both the SEE and $r^2 \times 100$ can be acceptable, the bias also needs to be considered. A technique can have low error and a good relationship but have a large bias, which makes it unacceptable. Bland–Altman analyses demonstrate bias and are discussed later in the chapter. Acceptable methods have a high $r^2 \times 100$, low SEEs, and no or low bias.

The line of best fit (regression line) is a slope and intercept that best minimize the squared deviations (difference between predicted and actual values) from the regression line. The SEE can be calculated from the correlation between methods (equation 1.5)

$$SEE = S_y \sqrt{1 - r^2} \tag{1.5}$$

where S_y is the SD of Y and r is the correlation between X and Y.

The SEE can also be calculated from the sum of the squared deviators from the regression line (equation 1.6)

$$SEE = \sqrt{\frac{\sum (y_i - \hat{y})^2}{n - 2}} \tag{1.6}$$

where y_i is the actual value (method a, reference method) for a given person and $\hat{y}$ is the point on the regression line predicted from method b (field method), using the line of best fit.

The differences between the predicted and actual values averaged over the entire sample (deviations squared) give an estimate of the degree of relationship between two methods independent of sample variability. Both the correlation and SEE assume a linear, or straight-line, relation between two variables. If the relationship is nonlinear, it will be underestimated.

Validation and Cross-Validation Studies

In the first part of this chapter, we presented key concepts related to assessing the validity of a given method of estimating body composition. Methods also need sufficient precision and objectivity to be valid; these errors can be quantified using the equations described (see table 1.4). In judging the usefulness of a new method, a well-designed validation study uses a reference method as a criterion for accuracy. Many approaches to the validation of a new method use correlation and regression analysis. Both approaches are useful but insufficient. The need for cross-validation of a new method or equation using the Bland–Altman approach is critical for complete evaluation. Using the Bland–Altman approach allows for the assessment of both systematic and random errors. Also, this approach can identify biases in the relationship at the ends of the distribution.

Thus, testing a new method against a reference method (the criterion method) is the first step in a validation study. An ideal reference method is the four-component

Table 1.4   Summary of Calculations for Assessing Various Measurement Errors in Body Composition

Equation or analysis	Use	Advantages or limitations	Interpretation
TEM	Gives the precision or objectivity	Result is in the unit of measure of the component measured	A smaller TEM means better precision or objectivity
ICC	Used to quantify reliability and is the relationship between repeated measures	A unitless number that allows comparison across components	Zero means unreliable, and 1 means perfectly reliable
SEM	Calculated using the ICC	Expressed in the unit of measure of the technique	The smaller the SEM, the more reliable
RMSD	Gives the accuracy of a technique	Expressed in the unit of measure of the technique	The smaller the RMSD, the more accurate
Correlation/ regression	Used extensively in validation studies; gives the relationship between the new method and the criterion method	Can underestimate the relationship for homogeneous samples	Ranges from 0 to 1 and can be positive or negative; the closer to 1 (or −1), the higher the relationship
SEE	Gives the standard deviation of a regression line developed from a validation study	Provides a better estimate of relationship than correlation	The smaller the SEE, the more accurate: 2%-3% for level 1 methods and 3%-4% for level 2 methods
Bland–Altman	Plots the difference between two methods (y-axis) against the reference value (x-axis) in a validation study	Necessary step in a validation or cross-validation study; can show a systematic error or random error across the full range of values	For accuracy, the slope should be 0 and the mean difference should be 0

TEM = technical error of measurement; ICC = intraclass correlation coefficient; SEM = standard error of measurement; RMSD = root mean squared deviation; SEE = standard error of estimate.

model using body density, body water, and bone mineral measured from laboratory methods for each subject to estimate body fat (discussed in depth in chapter 2). The SEE can be used to estimate the accuracy of the new method. Depending on the size of the SEE, we can classify the new method as a level 2 (SEE = 2%-3%), level 3 (SEE = 3%-4%), or level 4 (SEE = >4%). We can use this approach of assigning levels only when we use an established reference method, such as the four-component model, as the criterion.

After a body composition equation has been developed from a validation study in a given population, the equation can be applied to another sample of the same population to test how well the equation works in the new sample or in a new population. Cross-validation is important because the process ensures that a new technique is not specific to one laboratory or to one population of people. If we are cross-validating an equation, the results obtained for the RMSD should be similar to the SEE of the equation in the sample in which it is developed. Investigator- and investigation-specific equations are those equations that have been shown to work poorly on another sample of the same population because of investigation differences in protocol as well as sampling differences between studies. Population-specific equations are those equations that have been tested on samples from one population and have been shown to work poorly on another population. Thus, if you apply an equation that works well for adults but not for children, you will get a systematic difference between the reference method and the new method applied to the wrong population.

In a cross-validation study, if the mean predicted value is close to the actual value and the SEE is similar to the original validation study, we have evidence that the equation has been cross-validated. Combining the systematic error (mean difference between methods) and the SEE yields the total error for the new method and equation (RMSD). Additional cross-validation studies can be done in other populations to see whether the method works as well in the new population.

Another aspect of validation and cross-validation is to be sure that accuracy is not affected by the magnitude of the value being measured. For example, it is possible that a technique is valid in normal-weight populations but not in overweight populations. Bland–Altman analysis (6) allows you to determine whether the difference between the new method and the reference method increases (or decreases) as the values being measured increase or decrease. In this approach, the new method is graphed on the *x*-axis against the difference between the reference method and the new method for each subject (*y*-axis). This approach is a better way to assess cross-validation results than the RMSD approach because it separates out the random error from the systematic error.

In the ideal case (figure 1.1), the slope of the line of best fit and the mean difference between methods are both zero. This indicates that there is no systematic error between methods (zero mean difference) and no relation in the difference between methods and the absolute values of fatness (zero slope).

In the case where there is a relationship between methods dependent on the absolute value of fatness, we illustrate two hypothetical results. In figure 1.2, although the slope of the best fit line is still zero, there is an average systematic error of +3% compared to the reference method at all measured body composition values.

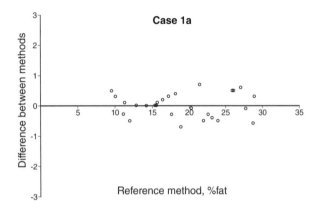

FIGURE 1.1 Bland–Altman comparison of two methods where the difference is zero.

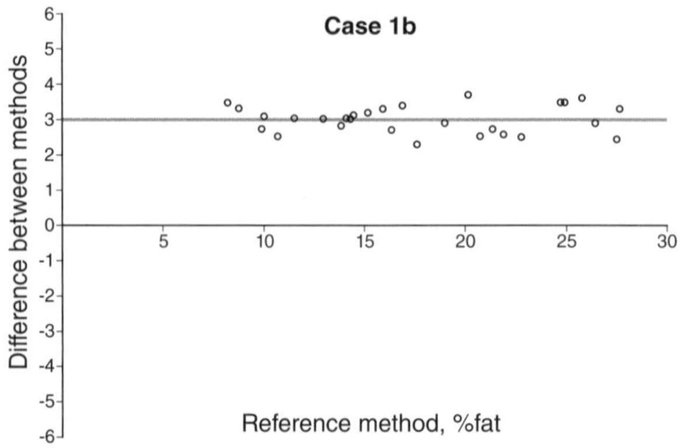

FIGURE 1.2 Bland–Altman comparison of two methods where the difference between methods is +3.

In figure 1.3, we can see that at low body fat values for the reference method, the differences between methods are all below the line, indicating that the new method produces smaller values than the reference method and thus underestimates body fatness. At high values of body fat by the reference method, the new method overestimates fatness. The opposite is shown in figure 1.4.

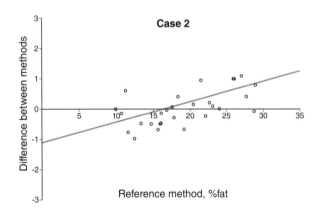

FIGURE 1.3 Bland–Altman comparisons between methods where the difference between methods is positively related to the absolute value of body fatness.

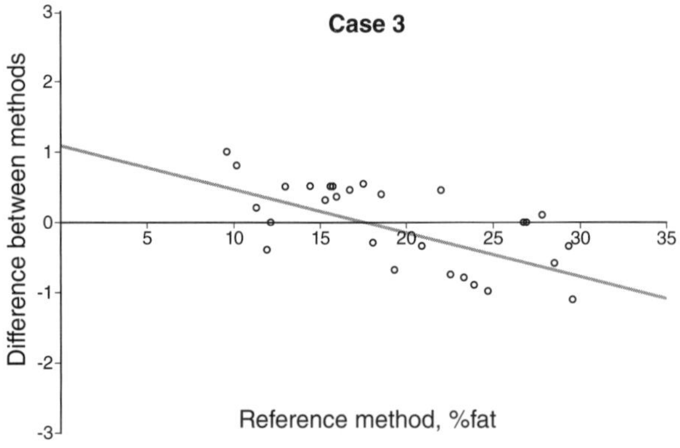

FIGURE 1.4 Bland–Altman comparison of two methods where the difference between methods is related negatively to the absolute value of body fatness.

Body Composition Terms and Concepts

Though many people think of body composition as the measurement of fatness, body composition is a broad field that includes the measurement of any portion of the human body. Therefore, there are many terms that need to be explained. Also, the body composition methods that have been developed are based on various ways of summing the parts that make up the whole body. In this section, we define terms and concepts related to body composition as well as introduce models that are commonly used.

- *Two-component model.* Considered to be the simplest and most easily applied body composition model, using only FM and fat-free mass (FFM) to estimate body mass
- *Multicomponent model.* Divides the body into three or more components (discussed in detail in chapter 2)
- *Fat mass.* All extractable lipids from adipose and other tissues in the body
- *Lean body mass.* Collective weight of the internal organs, bones, and muscles plus essential fat in the organs, central nervous system, and bone marrow
- *Mass.* The amount of matter contained within the body
- *Weight.* The amount of force acting on the body from gravity
- *Body mass (total body mass).* The sum of the individual mass of each of the constituents or components that make up the body
- *Bone mass.* The mass of all bone in the body

Then, we distinguish between adipose tissue and fat.

- *Adipose tissue.* Connective tissue consisting of adipocytes, which are cells that store mainly triglycerides; adipose tissue varies in fat content from 60% to 85% with a protein and water content of 2% to 10% and 10% to 20%, respectively

Next, we compare FFM with lean body mass and lean soft tissue (see figure 1.5).

- *Lean body mass.* The FFM plus essential lipids (Behnke first proposed the concept of essential fat [7].)
- *Lean soft tissue.* Consists of water and protein (muscle: skeletal, cardiac, smooth) and organ tissue but excludes mineral (FFM [no lipids] is always less than lean body mass, and lean body mass is always greater than lean soft tissue mass.)
- *Fat-free mass (also fat-free body).* All lipid-free tissues in the body

Several body composition methods utilize the concept of body density rather than body mass.

- *Density.* Mass per unit volume (density = mass/volume) (mass units: milligrams, grams, or kilograms; volume units: milliliters, liters, cubic centimeters, cubic meters, and so on)
- *Body density.* Overall density resulting from the density of the major molecular components of the body, such as fat, minerals, water, and protein
- *Densitometry.* The method of estimating body composition by using the known or measured densities of various body composition components
- *Bone mineral density.* The density of bone after all water has been removed, which is approximately 3.038 g/cm^3 (0.110 $lb/in.^3$) and represents the greatest density of all tissues in the body

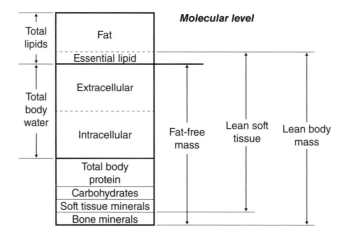

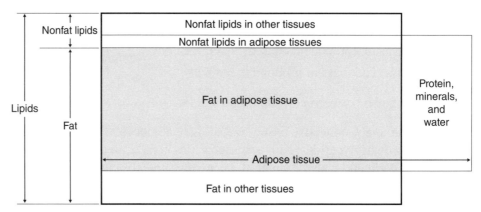

FIGURE 1.5 Main components of the molecular level of body composition.

Reprinted by permission from W. Shen et al., "Study of Body Composition: An Overview," in *Human Body Composition*, 2nd ed., edited by S. Heymsfield, T. Lohman, Z. Wang, and S. Going (Champaign, IL: Human Kinetics, 2005), 1-16.

- *Density of fat.* The lowest density of any tissue in the human body at approximately 0.9 g/cm³ (0.03 lb/in.³) (Here fat is lipid.)
- *Density of protein.* A primary constituent of all tissue in the body with an approximate density of 1.34 g/cm³ (0.0484 lb/in.³)
- *Density of water.* Has a density of 0.9937 g/cm³ (0.0359 lb/in.³) at standard temperature and pressure

Now we deal with components of the fat-free body.

- *Body water (total body water).* The amount of water in the body is often misquoted; however, context is the key. If we consider the water portion of FFM, water makes up approximately 74% of the FFM. If we consider the water portion of the entire body, water makes up approximately 55% to 60% of the total body. Total body water can vary considerably depending on hydration status.
- *Protein mass.* Mass of all the protein in the body
- *Mineral mass.* Mass of all the mineral in the body (includes osseous and non-osseous mineral)

- *Dry fat-free mass*. Composed of protein and mineral that has had the water removed
- *Models*. The basis from which the estimation of body composition arises; can consist of as few as 2 to as many as 11 or more components, depending on the level of analysis used to create the model

Finally, we define reference man and reference child. Reference man is the theoretical human model developed in the 1960s that gave rise to all modern body composition research. Reference man is Caucasian, weighs 70 kg (154 lb), and is composed of 15% FM (10.5 kg [23.1 lb]) and 85% FFM (59.5 kg [131.2 lb]) with a 6.8% mineral, 19.4% protein, and 73.8% water content (8). Other reference models have been created, including reference woman (10) and reference child with a higher water and lower mineral content of the FFM as compared to reference man (9).

Summary

The first part of this chapter introduced four levels of accuracy for classifying different body composition methods followed by an overview of all chapters.

The second part of this chapter introduced the concepts of quantifying measurement errors, including precision, objectivity, reliability, and accuracy. Also, the TEM was defined and illustrated. The validation and cross-validation studies were explained as essential parts of developing a new method or new equation.

The last part of the chapter introduced key body composition end points and concepts that need to be mastered before chapter 2 is presented. Chapter 2 will present a more detailed discussion of the pertinent body composition models and how they relate to the measurement methods used today.

In summary, the goal of this text is to describe the available body composition techniques and their theoretical underpinnings, distinguish between methods that are laboratory based versus field based, present assessment tips to improve measurement accuracy and consistency, and present real-world applications of body composition measurements. The chapters that follow will increase your appreciation of the complexity of body composition assessment and help you to perform these measurements with accuracy and precision.

2

Body Composition Models and Reference Methods

Jennifer W. Bea, PhD
Kirk Cureton, PhD, FACSM
Vinson Lee, MS
Laurie A. Milliken, PhD, FACSM

LEARNING OBJECTIVES

After completing this chapter, you will be able to do the following:

- Describe the levels of human body composition assessment based on their accuracy
- Describe the models of human body composition upon which reference methods of assessment are based
- Describe the limitations and accuracy of the reference methods used to assess body composition

Methods to assess body composition vary in validity, applicability, and expense, but all methods exploit certain characteristics that are common to all humans. The purpose of this chapter is to first describe the models of body composition and then describe the reference methods upon which those models are based. Body composition models are simply different ways that the human body can be viewed; they are different methods by which we can sum the parts to equal the whole (i.e., body weight). The methods developed to measure the composition of the whole body focus on the direct measurement of each part or many of the parts, the measurement of some substance that relates to the amount of a part, or a combination of both. The reader will learn about these reference methods, how they work, how they relate to the models of body composition, and why they are the most accurate methods available.

The first comprehensive discussion of body composition models was published by Wang et al. (1), despite research having been conducted in this area since the 1800s. Prior to the work of Wang et al., the two- and three-component models dominated the field. The two-component model allows for the measurement of one component and, given certain assumptions, for the indirect estimate of the remainder. For instance, during the 1960s and 1970s, body densitometry was commonly used to estimate percent body fat assuming the density of the fat-free body was 1.1 g/cm^3 (0.04 lb/in.^{3}) and the density of fat was 0.90 g/cm^3 (0.33 lb/in.^{3}). It was often called the gold standard. However, body water and bone mineral variation affect the density in the fat-free body considerably, thus leading to a three-component and four-component system, where both body water and body density or body water, body density, and bone mineral can be assessed. Selinger (2) was the first to develop a four-component model expanding the early work by Siri (10) using the three-component model. Lohman (3) applied Selinger's four-component model to children who have a higher water content and lower bone mineral content in their fat-free mass (FFM). This work is further described in chapter 3.

Levels of Human Body Composition

Wang et al. (1) described five levels of body composition models: the atomic, molecular, cellular, tissue-system, and whole-body levels. Wang et al. also sought to clarify and standardize terminology related to body composition so that researchers could further develop accurate assessment methods, which rely on a common understanding of body composition models. This chapter reflects these standard terms and allows for a more sophisticated approach to the field of body composition assessment.

Atomic Level

The atomic level of describing the human body relies on the evidence that humans are composed of 50 atoms, or elements, of which 6 comprise more than 98% of total body weight (4). Because a small number of atoms explains the majority of body weight, this model includes the 11 most abundant atoms, which comprise over 99.5% of the body. These 11 atoms are oxygen, carbon, hydrogen, nitrogen, calcium, phosphorus, sulfur, potassium, sodium, chlorine, and magnesium. The remaining 39 atoms account for less than 0.2% of body weight and are called the residual mass (R) in this model.

The equation for this model is

$$\text{Body weight} = O + C + H + N + Ca + P + S + K + Na + Cl + Mg + R \qquad \textbf{(2.1)}$$

where each element is abbreviated using its periodic table symbol and R is residual mass (1).

Although reference data for equation 2.1 comes from cadaver analysis or biopsy, in vivo measurements can also be made for most of the major elements in this model using total body potassium (TBK) counting and neutron activation analysis (NAA) for sodium, chlorine, nitrogen, and carbon, which together account for 98% of total body weight. Both approaches will be discussed in more detail later in this chapter.

Molecular Level

The next level is the molecular level where the 11 major elements described in the preceding section are combined to form molecules. Although the human body contains thousands of molecules, which would be nearly impossible to measure individually, researchers studying this level of body composition have grouped together closely related molecules to facilitate their measurement. For this reason, there can be many models that reflect the molecular level, depending on what measurement methods are available. For example, water (aqueous), lipid, protein, mineral, and glucose account for more than 98% of body weight (4), but each of these molecules is found in several body components and tissues. This makes their measurement difficult because there may not be one method to measure any given molecule across all of its different tissues. Lipids illustrate this assessment complexity. Lipids can be simple lipids, compound lipids, steroids, fatty acids, and terpenes and can be categorized as essential or nonessential. Essential lipids can be found in cell membranes, whereas nonessential lipids, such as triglycerides, can be stored in adipose tissue. Nonessential lipids are also known as fats. Therefore, a method that can measure fat may not also measure the essential lipids present in cell membranes. The molecular equation in this case is

$$\text{Body weight} = \text{lipid} + \text{aqueous} + \text{protein} + \text{mineral} + \text{glucose} + \text{residual} \qquad \textbf{(2.2)}$$

where residual includes compounds not in the previous categories and in amounts that are 1% of body weight (1).

There are accurate direct measurement methods for the aqueous and mineral components of the molecular model. Total body water (TBW) can be measured by isotope dilution techniques, such as deuterium and tritium dilution, and total body mineral (TBM) can be measured by dual-energy X-ray absorptiometry (DXA). Proteins, which are molecules containing nitrogen, can be estimated indirectly by measuring the total body nitrogen (TBN) content using NAA. This assumes that all nitrogen in the body is in the form of a protein and that the percentage of protein that is nitrogen is known (16%) and is constant among people (5). Total body fat can be measured by total body densitometry where fat is assumed to be a density of 0.90 g/cm^3 (0.33 $lb/in.^3$) and FFM is assumed to be 1.1 g/cm^3 (0.04 $lb/in.^3$) (6,7).

Other equations on the molecular level are (1)

$$\text{Body weight} = \text{aqueous} + \text{dry body weight} \qquad \textbf{(2.3)}$$

where dry body weight is the sum of lipid, protein, mineral, glucose, and residual;

$$\text{Body weight} = \text{lipid} + \text{lipid-free body weight} \tag{2.4}$$

where lipid-free body weight is the sum of aqueous, protein, mineral, glucose, and residual;

$$\text{Body weight} = \text{fat} + \text{LBM} \tag{2.5}$$

where fat is nonessential lipids and lean body mass (LBM) is the sum of essential lipids, aqueous, protein, mineral, glucose, and residual.

Cellular Level

The cellular level reflects that molecules are arranged into cells that perform coordinated functions. As was the case in the molecular level, it is not possible to measure every cell, so researchers have grouped cells into those that have similar functions or attributes, which results in four groups: connective, muscular, nervous, and epithelial cells. Adipocytes, blood cells, and bone cells are categorized as connective, and all striated skeletal, smooth, and cardiac cells are considered muscular. Also included in the cellular level are extracellular fluids (ECF) and extracellular solids (ECS). The cellular equation is

$$\text{Body weight} = \text{cell mass} + \text{ECF} + \text{ECS (1)} \tag{2.6}$$

where cell mass includes all four cell groups, ECF is the sum of plasma volume and interstitial fluid, and ECS is the sum of inorganic and organic solids.

Because there is no one method for measuring cell mass, equation 2.6 has been modified to reflect the components that can be accurately measured. The equation becomes

$$\text{Body weight} = \text{fat cells} + \text{body cell mass} + \text{ECF} + \text{ECS} \tag{2.7}$$

where body cell mass (BCM) includes cells that are energy metabolizing, which includes the protoplasm of fat cells but not the triglyceride stored within them, and ECF and ECS, as described previously (1).

BCM cannot be measured directly but can be estimated by TBK counting using an equation (BCM = $0.00833 \times$ TBK) (8). ECS also cannot be measured directly; however, it can be estimated by measuring certain ECS, such as total body calcium (TBCa), through NAA, where ECS = TBCa/0.177 (7). The plasma volume and interstitial fluid portions of ECF can be measured by dilution techniques.

Tissue-System Level

Because cells organize themselves into tissues, the body can also be defined by the tissues, organs, and systems that have related functions. Like the four categories of cells, there are four categories of tissues: muscular, connective, epithelial, and nervous. Three tissues—bone, adipose, and muscular—are important foci in health and wellness research; these tissues account for about 75% of total body weight (4). A tissue-level model is

$$\text{Body weight = muscular tissue + connective tissue} \\ \text{+ epithelial tissue + nervous tissue (1)} \qquad \textbf{(2.8)}$$

Another tissue-system level model is based on the nine main systems present in the body. This model is

$$\text{Body weight = musculoskeletal + skin + nervous + circulatory + respiratory} \\ \text{+ digestive + urinary + endocrine + reproductive system (1)} \qquad \textbf{(2.9)}$$

Although the tissues and systems in this model are important to health and wellness professionals, the use of this model is rather limited because currently, there are few methods available to measure each of these tissues or systems. Therefore, a more useful approach has been to divide these tissues and systems into groupings where measurement methods are available. This has resulted in the following equation, which includes 85% of body weight with 15% explained by the residual (4):

$$\text{Body weight = adipose tissue + skeletal muscle + bone} \\ \text{+ viscera + blood + residual (1)} \qquad \textbf{(2.10)}$$

The most complete information about this model has been derived from cadaver analyses. However, in vivo techniques are available as well. Adipose tissue can be estimated by densitometry, computed tomography (CT), or magnetic resonance imaging (MRI), and bone tissue can be estimated by DXA. Skeletal muscle can be estimated indirectly from 24-h urinary creatinine excretion, TBK, or NAA for nitrogen. Appendicular skeletal muscle can be estimated by DXA from lean tissue mass and valid equations that have been developed. These techniques are discussed in subsequent sections.

Whole-Body Level

The whole-body level includes measurements that describe the external characteristics of the body as a whole, such as size and shape. Lohman et al. (9) described 10 of these external characteristics: stature, body weight, body mass index (BMI), segment lengths, body breadths, circumferences, skinfold thicknesses, body surface area, body density, and body volume.

Stature reflects body size and is often used to characterize growth in children. Body weight is also used as a screening tool during childhood growth and during adulthood to monitor under- and overweight. BMI is body weight/height^2 and is also used to monitor obesity in both youth and adults. Another index with height and weight is called the Fels index, which is body weight$^{1.2}$/height$^{3.3}$, which correlates with fatness slightly better than BMI (9). Other measures of body shape and roundness are also available as indexes of body composition (see chapter 4). Body surface area is often used to estimate basal metabolic rate when measures of FFM are not available.

More specific dimensions of the body can be measured with body segment lengths, breadths, circumferences, and skinfold thicknesses. The most common segment lengths are upper and lower extremity lengths and sitting stature. Body breadths are used to estimate frame size and skeletal mass. Commonly measured breadths are wrist, elbow, ankle, knee, and bi-iliac. Circumferences are used in equations predicting percent body fat and are especially useful in obese populations. The waist circumference has been widely used as a measure of abdominal fatness, which is associated with a higher risk of cardiovascular diseases. Skinfold thicknesses involve measuring the subcutaneous fat at specific standardized anatomical locations. Many prediction equations exist to estimate percent body fat from three to seven skinfold thicknesses.

Body volume and body density are the final whole-body measurements. Body volume is a measure of body size and is also used in the calculation of body density, which is body weight/body volume. Body density can then be used to estimate body fatness (called densitometry), assuming that the densities of fat and FFM are known and stable between and within people (10). Several population-specific values have been determined for the density of FFM, which has improved the prediction of percent body fat.

PRACTICAL INSIGHTS

Different levels of body composition may be more pertinent to one field than to another given the focus of the field. For example, in nutrition, the molecular level might be more useful since the equation includes lipid, protein, and glucose. In exercise science, the tissue-system level is useful since adipose tissue, skeletal muscle, and bone are in this equation. In epidemiology, the whole-body level measurement of body mass index is frequently used. Your doctor may focus on the whole-body level when assessing your height and weight. This text primarily focuses on methods used to measure the components of the levels that are often used in health-related fields.

Models of Human Body Composition

Multicomponent models are one of the best reference methods for in vivo estimation of body composition. We call them level 1 methods for assessment of body fatness. Although the most accurate compared to other methods, multicomponent models involve several techniques, each containing technical errors and assumptions to estimate body composition. However, the assumptions of multicomponent models are fewer than simpler models. Each of the components will be discussed in the following sections, including methods of measurement and equations for estimating body composition.

Two-Component Model

Many techniques for estimating body composition have been based on a two-component chemical model, in which the body mass (BM) is assumed to be made up of only two components, fat and FFM. The fat mass (FM) includes all ether-extractable lipid in the body; the FFM includes all other tissues. For example, a two-component model is assumed in estimating body composition from the measurement of body density using underwater weighing (hydrodensitometry) or air displacement plethysmography. This technique used to be considered the gold standard of indirect methods for estimating body composition, and most field methods, such as estimations of body composition from skinfold thickness or other anthropometric measures or from bioelectrical impedance analysis (BIA), have been validated against this laboratory measure based on a two-component model (11).

In estimating body composition using a two-component model, the composition (makeup) of the two components is assumed to be a constant percentage of the FFM from one person to another. For example, in estimating body fatness from body density determined by underwater weighing in adults, it is usually assumed that the density of fat is 0.90 g/cm^3 (0.33 lb/in.^{3}) and the density of the FFM is 1.1 g/cm^3 (0.04 lb/in.^{3}) and that the FFM comprises 73.8% water, 19.4% protein, and 6.8% mineral. In estimating body composition from the measurement of TBW by the dilution technique, it is assumed that water makes up a constant fraction of the FFM (10,12).

The primary limitation of the two-component model is the assumption that the makeup of the FFM or the concentration of a substance within the FFM is constant. Although the methods for estimating body composition based on a two-component model are reasonably accurate in most people, variability among individuals in the water, protein, and mineral fractions of the FFM and in concentrations of other constituents of the FFM results in considerable error in estimates of body composition in individuals who do not conform to the assumptions (13). In addition, systematic errors exist in some population subgroups, such as children (14), diseased individuals (15), and some groups of athletes (16,17), in which the makeup of the FFM is different than assumed. In these groups, different constants must be used for accurate estimations of body composition using a two-component model.

PRACTICAL INSIGHTS

Using body composition methods that rely on the two-component model has limitations due to the systematic error that may be present. Percent body fat can be under- or overestimated by 2% to 8% fat in a subgroup where the assumptions present in the two-component model are violated. Often, the technician will not know exactly by how much their values are off. It is important to realize that no body composition technique is error free and that error is propagated, meaning it accumulates. Error in the choice of method is added to any error present in how the method is carried out. A well-educated technician should be able to choose the most appropriate method that has the lowest error for the person being measured and for the purpose of the measurement. Then the protocol should be carefully carried out to produce the lowest measurement errors for the method chosen to produce the most accurate result.

Three-Component Model

To reduce assumptions included in the two-component model, other multicomponent models have been developed. The three-component model is similar to the two-component model but includes an additional estimate of some component of FFM, such as TBW (i.e., fat, water, residual), using tritium- or deuterium-labeled water dilution techniques. The measurable components are body weight, TBW, and body volume to derive estimates of FM, TBW, and dry FFM (mineral and protein). Estimation of body composition from this technique assumes specific hydration levels (10). Other three-component models exist where other components of FFM are measured directly.

Four-Component Model

The four-component model builds on the three-component model by adding an estimate of mineral (i.e., fat, water, mineral, and residual). The measurable quantities are BM, body volume, TBW, and mineral. BM is measured on a scale. Body volume is measured by underwater weighing, air displacement plethysmography, or in vivo neutron activation (IVNA) analysis. TBW is typically derived from tritium- or deuterium-labeled water dilution, and bone mineral (Mo) is derived via DXA. The equation is as follows:

$$BM = FM + TBW + Mo + residual \qquad (2.11)$$

Several equations have been derived to predict body composition from four-component models (18), particularly FM. The most recent improvements in estimates have been for children and adolescents and lean soft tissue (19,20).

Alternatively, the four-component model to estimate FM could be computed by measuring BM, TBW, total body protein (TBPro = TBN × 6.25), and total body bone

---

PRACTICAL INSIGHTS

Four-component models have tremendous advantages over other laboratory methods because they combine the strengths of several low-error methods to give an overall accurate estimate of body composition. However, it is important to stress that an appropriate four-component model is one that utilizes laboratory methods to measure the four components. There is a growing trend for some laboratories to use field methods to measure some of these components. When this is done, the level of error increases, making the overall accuracy unacceptable. Therefore, a four-component model should be performed following the standardized protocols for densitometry (for body volume), dual-energy X-ray absorptiometry (for bone mineral), and tritium- or deuterium-labeled water dilution (for total body water) laboratory methods. This will ensure the lowest error rates for the overall estimate of body fatness.

ash (TBA = TBCa/0.34) to estimate the four components of fat, water, protein, and bone ash for the following equation:

$$FM = BM - (TBW + TBPro + TBA) \qquad (2.12)$$

The four-component model has significant advantages over other models and measurement techniques during puberty because assumptions in less robust models may be violated as tissues mature (14,18). The four-component model is an excellent reference method for validating field measures and new body composition methods as well as other laboratory techniques.

Five-Component Model

In an attempt to improve on the four-component model, a five-component model in which the body is conceived as consisting of water, protein, mineral, glycogen, and fat has been used to measure body composition in healthy and diseased individuals (15). In applying this model, NAA is used to measure TBPro, and tritium dilution is used to measure TBW. TBM and TBPro are estimated by assuming a constant relationship to TBW. The sum of the first four components is the FFM. Fat is obtained by the difference of BM and FFM. The advantage of this model compared to two-, three-, and four-component models is that TBPro is directly measured. The limitation is that body mineral and glycogen are not directly measured.

Six-Component Model

In the search for a more refined, definitive reference method that could be used with living humans, a six-component chemical model of body composition was proposed (5) in which the body is conceived of consisting of water, protein, glycogen, osseous bone mineral, non-osseous bone mineral, and fat. By using neutron activation methods to measure total body nitrogen, calcium, chloride, sodium, and carbon; whole-body potassium-40 counting to measure total body potassium; and tritium dilution to measure TBW, each of the components in the model was measured directly, accounting for over 97.5% of body weight. This model has the advantage over the four- and five-component models in that total body fat is directly measured from total body carbon rather than calculated as the difference between body weight and measured components. High correlations existed between actual body weight and body weight calculated from the six-component chemical model. A strong relationship also existed between measured body density and body density calculated from the six-component chemical model. These findings suggest that this approach was very accurate (5). Using a slight modification of the approach, six-component model estimates of total body fat were compared with estimates from 16 other laboratory and field methods (20). Very high agreement was obtained between estimates from the six-component model and estimates from three- and four-component models that included measurement of TBW. There was slightly less agreement with estimates from DXA based on a three-component model and underwater weighing using a two-component model. The six-component model appears to be the most accurate method for measuring body composition in living humans; however, access to the appropriate measurement techniques limits its applicability.

Total Body Potassium Counting and Neutron Activation Analysis

TBK counting and NAA provide elements for use in atomic and multicomponent models of body composition described in the preceding section. Potassium is distributed solely in FFM, with a large proportion of it in skeletal muscle mass. Equations can be used to estimate BCM (cellular components of muscle, visceral organs, blood, and brain), FFM, or skeletal muscle mass from TBK measured in vivo by a whole-body counter (22).

In NAA or IVNA analysis, a fast beam of neutrons is applied to a person by an external source. This process excites atoms in the body, and more gamma rays are given off than would normally be given off as a result of decay of naturally occurring radioactive isotopes for particular elements in the body. The gamma rays are detected by a whole-body counter. The energy level of the gamma rays is used to identify the elements of origin; each element has a signature energy. The activity level indicates the abundance of the element. The elements that can currently be measured in vivo in humans are potassium, calcium, sodium, chlorine, phosphorus, nitrogen, hydrogen, oxygen, carbon, cadmium, mercury, iron, iodine, aluminum, boron, lithium, and silicon. These elements can be measured accurately, provided individuals remove street clothing, jewelry (including watches), and eyeglasses to reduce measurement contamination (22).

Although access to whole-body counters to measure NAA-related gamma emissions is now limited (fewer than six remain operational), they were once abundant. Building a specially shielded room for either TBK counting or NAA and properly equipping it is likely cost prohibitive today (up to USD 300,000), especially in light of the access to and cost of alternative technologies. However, understanding TBK and NAA is important for the interpretation of historical literature where direct assessment of body composition has been made. Additionally, measurements can be performed in existing facilities with sensitive high-energy gamma ray detectors, sufficient background radiation shielding, and quality control procedures (i.e., regular phantom testing).

Imaging Methods

Imaging methods, such as CT and MRI, are currently the reference methods with the greatest accuracy for in vivo quantification of body composition. They involve higher costs but are necessary for validating other methods. CT and MRI involve different technologies and advantages, as described in the following sections.

Magnetic Resonance Imaging

MRI is an in vivo measurement approach that quantifies body composition at the tissue-system level (23). The tissue-system level consists of skeletal muscle, adipose tissue, bone, blood, and organs/organ systems (see figure 2.1).

MRI takes advantage of the element hydrogen (H), which is one of the most abundant unbound elements found in large quantities in all biological tissue (23). Hydrogen nuclei (protons) have high affinity for alignment with a magnetic field and

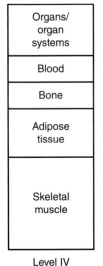

| Organs/ organ systems |
| Blood |
| Bone |
| Adipose tissue |
| Skeletal muscle |

Level IV
(tissue system)

FIGURE 2.1 The tissue-system level of analysis.

will interact and align or orient themselves in the direction of the strong magnetic field generated by the MRI scanner (23,24). Once aligned, a pulsed radio-frequency field is applied to the hydrogen protons. Large numbers of hydrogen protons, though not all, will absorb the energy from this field, and once the field is removed, the hydrogen protons release this stored energy as they return to their original state (relaxation). Acquiring images of different tissues is accomplished by manipulating the radio-frequency pulses to take advantage of the different relaxation times (T1) for each tissue of interest (23). Using the different relaxation times for lean and adipose tissues/fat, MRI can distinguish between and measure the two tissues. Tissue contrast can be increased by manipulating the time interval of the radio-frequency pulse and time to detect the induced signal within the hydrogen proton. This time is referred to as the echo (TE). The entire process of magnetic alignment, energizing the proton radio-frequency pulses, and measurement of the induced signal is often referred to as the pulse, or spin-echo, sequence (24). Signal intensity of the various tissues is the result of different T1 times and allows the discrimination between lean tissue, adipose tissue/fat, and other tissues. Tissues with a short T1 time appear bright in an MRI image, whereas tissues with a long T1 time will appear dark. Relaxation, or T1, time is the product of the arrangement of hydrogen protons, which differs in adipose tissue versus lean tissue. Thus, the signal intensity of adipose tissue is brighter than lean mass and organs in a T1 weighted image.

Acquisition of whole-body scans used to estimate body composition divides the body into appendicular and abdominal sections (see figure 2.2). The appendicular section can be further divided into lower (legs) and upper (arms) appendicular sections. Transverse (axial) "slices" are acquired, with the number of slices dependent on slice thickness and distance between slices. Total time for a whole-body scan depends on these two protocol settings. A greater number of slices provides greater precision and accuracy but increases scan time. In general, whole-body scans require approximately 30 min, depending on the scanner.

MRI is considered to be the reference method for validation of in vivo measurement of lean and adipose tissue by DXA, BIA, skinfolds, and anthropometry. MRI itself must be validated against direct methods of tissue measurement using human cadavers and deceased animals. The most commonly cited MRI validation studies for body composition (cross-sectional area, adipose tissue, and muscle tissue) are briefly summarized.

Engstrom et al. (26) used MRI to determine cross-sectional area of the thigh in three male cadavers using 10 mm (0.4 in.) slices in a serial and contiguous manner along a 40 cm (15.8 in.) section of the thigh. The image-based cross-sectional areas were within 7.5% of the matching cadaveric cross sections following dissection. Abate et al. (27) validated MRI measures of adipose tissue by performing MRI scans on unembalmed cadavers, two male and one female, and then dissecting the cadavers to excise and weigh the adipose tissue. The difference between the two methods was small and not clinically relevant (difference: 0.076 kg [0.168 lb]; 95% confidence interval: 0.005-0.147 kg [0.011-0.324 lb]). Validation of MRI-derived muscle measurements was accomplished by Mitsiopoulos et al. (28). The researchers performed multiple MRI measurements of the cross-sectional area in both the arm and leg and then measured

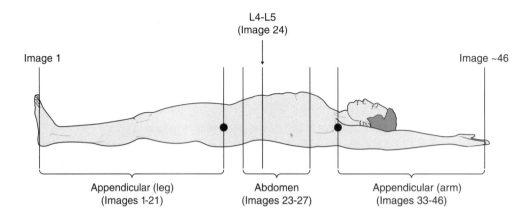

Protocol (abdomen)
T1-weighted, spin-echo pulse sequence
Each image = 10 mm thickness, 40 mm spaces
TR = 210 ms; TE = 17 ms; 1/2 NEX
FOV = 48 cm × 36 cm (rectangular)
Matrix = 256 × 256
Each acquisition = 7 images
Time = 26 (breath hold)

Protocol (appendicular)
T1-weighted, spin-echo pulse sequence
Each image = 10 mm thickness, 40 mm spaces
TR = 210 ms; TE = 17 ms; 1/2 NEX
FOV = 48 cm × 36 cm (rectangular) or 48 cm × 24 cm (1/2)
Matrix = 256 × 256
Each acquisition = 7 images

Sequence of series to acquire images
1. Sagittal scout to locate L4-L5 and right femoral head
2. L4-L5 down (abdomen protocol)
3. Femoral head down (appendicular protocol, rectangular FOV)
4. 35 cm below femoral head down (appendicular protocol, 1/2 FOV)
5. 70 cm below femoral head down (appendicular protocol, 1/2 FOV)
6. Sagittal scout to locate L4-L5
7. Coronal scout to locate right humeral head
8. L4-L5 up (abdomen protocol)
9. 35 cm above L4-L5 up (appendicular protocol, rectangular FOV)
10. Humeral head up (appendicular protocol, rectangular FOV)
11. 35 cm above humeral head up (appendicular protocol, 1/2 FOV)

Note: For Series 2 and 9, some of the images are discarded because they overlap with images in Series 3 and 10, respectively.

FIGURE 2.2   Commonly used scan protocol for body composition measurement using MRI. Specific settings may change depending on the MRI scanner used for measurement.

Reprinted by permission from R. Ross and I. Janssen, "Computed Tomography and Magnetic Resonance Imaging," in *Human Body Composition: Methods and Findings*, 2nd ed., edited by S. Heymsfield, T. Lohman, Z. Wang, and S. Going (Champaign, IL: Human Kinetics, 2005), 90.

the corresponding sites on the cadaver. The correlation coefficient between MRI and cadaver measures of appendicular muscle approached unity, an indication that the difference between the two measurements was essentially zero. Later, Hu et al. (28) performed multiple MRI scans on 97 excised porcine organ, muscle, adipose, and lean tissue samples immediately after necropsy ($N = 4$ pigs) and validated the measures by chemical analysis. The mean difference between MRI and chemical analyses was not significantly different from zero. These studies established the accuracy and precision of MRI for estimates of body composition (i.e., fat and muscle). MRI is not considered a reference method for the measurement of bone because bone lacks the abundant hydrogen atoms necessary for the MRI technique.

Notably, MRI does not use ionizing radiation like DXA, CT, or X-ray do (23), so there is no risk of exposure to radiation sources. However, MRI use is limited by the high cost of scanning, specialized training of the operator, risk of claustrophobia in some individuals, and the inability to measure large subjects because of the typical bore size of scanners. Additionally, any person with metal implants, metal stents, or nonremovable metal jewelry should not be measured using MRI techniques.

Computed Tomography

In CT scanning, both an X-ray tube and a receiver rotate perpendicularly around the body. The regions of interest may be selected in a fashion similar to that depicted in figure 2.2. The X-ray tube supplies ionizing radiation that is captured by the receiver after passing through the body and being attenuated by the different tissues the X-rays move through. The X-ray attenuation varies by characteristics particular to different tissues, relative to air and water, and each pixel derived from the attenuated X-rays is assigned Hounsfield units (HUs) to reconstruct images in grayscale for each image slice. To analyze the area of a given image, a technician may trace the perimeter of the target tissue or allow the software to identify the tissue by setting parameters for HU ranges. Once the area has been selected in either fashion, the area (in centimeters squared) is computed by multiplying the number of pixels by the surface area of the pixels. The volume of the tissue can be computed by integrating consecutive slices if available. The recommended model to derive tissue volume (V, in centimeters cubed) is the two-column model by Shen et al. (30), where t is the thickness of each individual image or slice, h is the distance between consecutive slices, and A_i is the area of the tissue (in centimeters squared) for the following equation:

$$V = (t+h)\sum_{i=1}^{N} A_i \qquad (2.13)$$

The tissue density may then be computed by using the constant densities of 0.90 g/cm³ (0.33 lb/in.³) for adipose tissue and 1.06 g/cm³ (0.038 lb/in.³) for skeletal muscle because of the relative homogeneity of the tissue densities among individuals (4,31,32). Skeletal muscle lipid content can also be estimated, but it is not to be confused with intra- or extramyocellular fat values obtained by biopsy. In a CT scan, a greater number of low-density skeletal muscle pixels equates to higher skeletal muscle lipid content. These estimates have been validated in adults and adolescents (33-36). Accuracy and reproducibility for various tissues have been determined by comparing images to direct measures of human cadavers, with correlation values ranging from 0.79 to 0.99, depending on the tissue (23).

CT precision is high, and CT scanners are relatively more common than MRI machines. However, there are several limitations. First, a significant exposure to ionizing radiation prevents widespread use of CT in body composition research. CT machines are also quite large and require adequate space for housing. Furthermore, a high level of technical expertise is required for accurate estimation of tissue composition.

Summary

In this chapter, five levels of body composition were presented along with the most common models of body composition assessment. CT, MRI, IVNA, and multicomponent models are the major reference methods, or gold standards, for accurate and precise body composition assessment. IVNA is no longer widely available in the United States. Both CT and MRI may be cost prohibitive for routine use but can be employed to validate any new methods to estimate body composition. Importantly,

CT and IVNA expose individuals to substantial radiation; MRI does not involve radiation. Alternative methods that have already been well validated against MRI and CT, such as DXA (37,38), involve dramatically lower or no radiation dose and lower costs (22).

3

Body Composition Laboratory Methods

Robert M. Blew, MS
Luis B. Sardinha, PhD
Laurie A. Milliken, PhD, FACSM

LEARNING OBJECTIVES

After completing this chapter, you will be able to do the following:

- Understand the theoretical basis for all six laboratory methods of whole-body composition assessment presented in this chapter

- Know the advantages and limitations of each body composition laboratory method

- Describe the accuracy, objectivity, and precision of each method using a standardized protocol

In this chapter, we cover the major laboratory methods of estimating body composition: densitometry, total body water (TBW), dual-energy X-ray absorptiometry (DXA), and ultrasound. Each of these methods provides an objective approach to the assessment of body composition with prediction errors of about 2% to 3% for percent body fat (level 2) accuracy. Each laboratory method is based on sound theoretical principles; however, certain assumptions limit their accuracy and use as reference methods where the two-component model is assumed.

Densitometry

Densitometry is a term used to refer to those measurement procedures that use total body density to predict body composition. It is based on the principle that the overall density of any given object is determined by the percentage and densities of its individual components. In other words, the density of a material is equal to the ratio of its mass and volume (equation 3.1).

$$\text{Density} = \text{Mass/Volume} \tag{3.1}$$

For determining the density of the human body (D_b), mass can be readily obtained by measuring the person's weight using a high-precision scale. Body volume, however, is a more complicated measurement to obtain, and thus the various methods used to assess volume fall into the category of densitometry.

In the classic two-component model of body composition, the mass of the body can be divided into two components, fat mass (FM) and fat-free mass (FFM). FM is considered to be a uniform component that contains all fat regardless of type (brown, white, subcutaneous, visceral, etc.), whereas FFM is the heterogeneous fraction comprising all nonfat material, including water, protein, and mineral. The model assumes a density for fat of 0.90 g/cm³ (0.33 lb/in.³) and a density for FFM of 1.1 g/cm³ (0.04

PRACTICAL INSIGHTS

Laboratory methods are not reference methods or gold standard methods. A reference method is one that new methods should be compared against to test for accuracy. Though laboratory methods have higher accuracy than less expensive methods such as field methods, they do not have a small enough error to be used as a gold standard against which to validate new methods. Validation studies for percent body fat should be performed using a multicomponent model (made up of laboratory methods) as the gold standard. This will ensure that the rate of error is the lowest possible and that the criterion percent fat will not vary between populations. If laboratory methods are used as a gold standard, more error will be introduced into the analysis, which will add to the error of the new method being tested. Remember that laboratory methods are 2% to 3% accurate while reference methods are 1% to 2% accurate.

lb/in.^{3}) and that they are constant across individuals regardless of age, gender, genetics, or health status. These density values are from early studies of animal carcasses and cadaver analysis in three males aged 25, 35, and 46 years (1,2). It was the knowledge of these values, however, that made in vivo study of body composition possible and enabled the development of percent body fat formulas from body density (D$_b$). Two of the simplest and most commonly used formulas are from Siri (3)

$$\%\text{Fat} = [(4.95/D_b) - 4.50] \times 100 \qquad (3.2)$$

and Brozek (1)

$$\%\text{Fat} = [(4.570/D_b) - 4.142] \times 100 \qquad (3.3)$$

The derivation of the Siri formula can be shown starting with the following equation:

$$\frac{1}{D_b} = \frac{f}{f_d} + \frac{\text{ffm}}{\text{ffm}_d} \qquad (3.4)$$

If we assume the density of fat is 0.90 g/mL (0.33 lb/in.^{3}), the density of fat-free mass (ffm$_d$) is 1.10 g/mL (0.04 lb/in.^{3}), f = fraction of the body as fat, and ffm = fraction of the body as fat-free mass, we can solve the equation for f to derive equation 3.2.

It is apparent that because these equations are based on body density, their validity is dependent on the assumption that the fractions and densities of the components are true for the individual being measured. However, the constituents that make up body fat, such as glycerides, sterols, phospholipids, and glycolipids, are less stable than other elements, and this can lead to varying levels of fat density from one individual to another and even within a given person when measured over time (4). More importantly, there is significant variability in the density of the FFM components because of changes in composition and density that are associated with growth and maturation, aging, and specialized training as well as differences based on gender and race. Even within a homogeneous population, there can be considerable variation among individuals. Estimates of the variation in fat-free body density are discussed in Lohman (5).

Estimation of Body Fat From Body Density

In the general population, Siri (3) has estimated that biological variation in water and mineral content of FFM limits the accuracy of body fat estimates from densitometry with a standard error of estimate (SEE) of 3.5% (5). It was further estimated that in a more homogeneous adult population, such as young adults, the SEE may be closer to 2.7% (table 3.1). This allows body density to be used as a criterion method for validation of a new method in selected populations where the water and mineral contents are known. In early studies with homogeneous populations, underwater weighing was used successfully as a reference method in validation studies of DXA (6,7).

In addition to biological variation, the most widely used methods of estimating body volume, underwater weighing (UWW) and air displacement plethysmography (ADP), also have inherent technical sources of variation. The following sections will consider these limitations to provide a better understanding of the merits and limitations inherent in all densitometric techniques. An additional, excellent overview of densitometry is covered in the review article by Going (8).

Table 3.1   Standard Deviations (SD) Associated With Biological Variation in Density as Composition of the Reference Body

Source of variation contributing to % fat and density	General population[1]		Specific population[2]	
	%Fat-SD	Density-SD (g/mL)	%Fat-SD	Density-SD (g/mL)
Water content	2.7	.0057	1.9	.0040
Protein/mineral ratio	2.1	.0046	1.5	.0033
Mean fat content of obesity tissue	1.8	.0039	1.3	.0028
Obesity tissue density	0.5	.0011	0.35	.0008
Mean fat content of reference man	0.5	.0011	0.35	.0008
Total	3.8	.0084	2.7	.0060

[1]From Siri (3) using estimates of variation components, as summarized by Keys and Brozek (2).

[2]Estimates in a specific population based on the assumption that each source is one-half the variation in the general population.

Reprinted by permission from T.G. Lohman, "Skinfolds and Body Density and Their Relation to Body Fatness: A Review," *Human Biology* 53, no. 2 (1981):181-225.

PRACTICAL INSIGHTS

For the two-component model to work well, the assumptions that the density of fat is equal to 0.9 g/mL and fat-free mass is equal to 1.1 g/mL need to be accurate. Variations from these reference values, such as those shown in table 3.1, will produce more error in percent fat values from the two-component model. While low SEEs in the prediction of percent fat can exist from a two-component model when measuring those whose fat and fat-free mass densities do not vary from the reference values, errors as high as 2% to 6% will exist when people vary in these densities. Most of the time, populations will vary in the components of fat-free mass. For example, bone density is lower in older populations. The water content of fat-free mass will vary in several populations (elderly and children) and over time (such as with weight loss and dehydration). In these cases, the SEEs associated with the two-component model will be inflated.

Underwater Weighing

UWW, also known as hydrodensitometry or hydrostatic weighing, was long considered the gold standard for body composition assessment. Although newer, more advanced techniques may have accentuated its limitations, it still serves as an acceptably precise and accurate method for most populations because of its fundamental principles. However, it is no longer classified as a gold standard (9). Ideally, UWW is performed in a laboratory setting in a specifically designed hydrostatic weighing tub or tank. The tank is generally constructed of redwood, stainless steel, or Plexiglas and is large enough to allow a large adult to be fully submersed without touching the sides of the tank. Typically, a chair assembly suspended from a spring-loaded autopsy scale or a platform supported on force transducers is situated centrally in the tank and provides the subject a place to sit, kneel, or lie prone. When the individual submerges completely, the scale or force transducers provide a measure of underwater weight.

Estimation of body volume by UWW is based on Archimedes' principle, which states that the upward buoyant force exerted on a body immersed in a fluid is equal to the weight of the fluid that the body displaces. In the case of UWW, when a person is submerged in water, body volume (V_{body}) is equal to the difference between body weight on land (W_{land}) and the weight in water (W_{water}), corrected for the density of water (D_{water}) as determined by its temperature at the time of submersion (equation 3.5). The body volume, along with body mass, is used to calculate body density (equation. 3.1) and then percent fat (equation 3.2 or 3.3).

$$V_{body} = (W_{land} - W_{water})/D_{water} \tag{3.5}$$

However, internal air or gas at the time of measurement, in particular gas in the gastrointestinal tract, air in the lungs, and air trapped by clothing, can falsely increase the measured body volume, so it is necessary to adjust for these factors. Although the volume of gastrointestinal (GI) gas can vary somewhat, the 100 mL (3.3 oz) proposed by Buskirk (10) has worked well as an estimate for the volume of gas in the gastrointestinal tract. The greater volume of air in the lungs, however, can vary considerably, significantly affecting the estimation of total body volume. Although the subject forcibly exhales all air from the lungs prior to the measurement, some air still remains, known as residual volume (RV). Incorporating RV and GI gas volumes, one can calculate body density (D_b).

$$D_b = W_{land}/[V_{body} - (RV + 0.100)] \tag{3.6}$$

In taller individuals, RV can contribute up to 2 L (0.5 gal) to the total body volume estimate, which, if unaccounted for, would yield highly inaccurate results. Thus, it is crucial to have an accurate RV estimate specific to the individual. The simplest means of predicting RV is by using body habitus, age, and gender, but even individuals matching in all three categories can have very different RVs. Therefore, it is best to have a measure of RV, preferably at the time of UWW.

There are two methods regularly used to measure RV. One is the closed-circuit technique and involves the dilution and eventual equilibration of an inert indicator gas, such as nitrogen, oxygen, or helium. In one such technique, oxygen dilution, the individual breathes in and out of a spirometer containing a known volume and concentration of oxygen until the concentration of nitrogen in the lungs and the spirometer equilibrate (11). A nitrogen gas analyzer measures the concentration in the flow of exhaled air. An initial estimate of nitrogen concentration is made after a maximal expiration, and this value is considered the initial alveolar nitrogen concentration (AN_{init}). Next, the subject takes five to eight breaths at two-thirds vital capacity from the spirometer until the nitrogen concentration reaches equilibrium (N_{equil}). A maximal inspiration and expiration are taken, after which the nitrogen concentration is measured again and recorded as the final alveolar concentration (AN_{final}). These values can then be applied to an equation that includes the volume of oxygen in the spirometer (Vol_{O2}), the nitrogen impurity present in the Vol_{O2} (N_{init}), the volume of dead space in the system (DS), and a factor correcting for ambient pressure, body temperature, and the spirometer temperature (BTPS), to determine residual volume.

$$RV = \left[\frac{Vol_{O2}(N_{equil} - N_{init})}{AN_{init} - AN_{final}} - DS \right] \times BTPS \qquad (3.7)$$

The other method is the open-circuit approach, where nitrogen is washed out of the lungs during a fixed period of oxygen breathing. It is the preferred method when measuring RV and underwater weight simultaneously because it can be quickly and easily performed. Flexible tubing connects to a mouthpiece inside the UWW tank so that the subject can perform the rebreathing procedure while submerged. The tube is connected on the outside of the tank to a breathing valve that can be switched between room air and a spirometer or a 5 L (1.3 gal) anesthesia bag connected to the oxygen. The subject submerges completely and performs a maximal expiration through the tube to room air. Underwater weight is obtained. The valve is then switched to the bag or spirometer, and breathing begins until nitrogen equilibrium is achieved.

Both approaches yield precise estimates of RV and, with appropriate equipment and procedural modification, can be used to measure RV either simultaneously with UWW (preferred) or outside the tank prior to measuring. Nevertheless, RV is the most significant contributor to error in the body density measurement. Using the law of propagation of errors, Akers and Buskirk (12) examined the effects of variations in body weight, underwater weight, water temperature, and RV on total body density. Results showed that RV is the major source of variation, with a change of just 0.1 L (3.4 oz) resulting in over a 0.5% change in body fat, whereas all sources combined were only 0.8% (12). Body density variation between 0.0015 and 0.0020 g/mL (0.000054 and 0.000072 lb/in.^{3}) is typical of the within-subject, trial-to-trial variation observed within a day (13). The technical error is somewhat larger for repeated estimates of body density over multiple days at 0.0030 g/mL (0.000108 lb/in.^{3}) (14), or 1.1% fat in men and 1.2% fat in women. The increased error between days is likely caused by fluctuations in body water and changes in GI gas (13). Food consumption or recent ingestion of carbonated drinks prior to measurement can change the estimated body fat by up to 1% (15,16). Similar effects are observed with body water changes, where dehydration or hyperhydration can cause up to a 2% error in body fat (16,17). Finally, estimating RV from height and weight rather than measuring it adds additional errors of 0.0030 g/mL (0.000108 lb/in.^{3}) and no longer qualifies densitometry as a laboratory method (18). As discussed in detail in chapter 2, biological variation in the components of FFM can vary with growth and maturation, aging, and specialized training, compromising the underlying assumption that the densities of FM and FFM are constant. Still, recognizing its relative affordability and how precise and accurate it is, UWW with RV directly measured is an excellent laboratory method for estimating body volume and the subsequent assessment of body composition.

Air Displacement Plethysmography

The general inconvenience of a water-based test coupled with the inability of many individuals to completely submerge after exhaling, whether because of a health condition or just a lack of confidence in water, can make achieving accurate body volume measurements by UWW difficult or even unrealistic. An alternative densitometric method that eliminates the challenges specific to being in water is air displacement plethysmography (ADP). Similar in principle to UWW, ADP is based on the theory that given a fixed, enclosed chamber, the volume of air displaced by any object placed within it will equal the volume of the object. Therefore, the body volume of a person can be estimated by sitting inside a sealed chamber and measuring the displaced volume of air. Body volume is indirectly calculated by subtracting the volume of air remaining in the chamber with the subject inside from the volume of air in the chamber when it is empty.

Although theoretically simple, early attempts to use ADP for the purpose of body composition assessment proved ineffective. Problems accounting for variability in the temperature, pressure, and humidity resulted in significant volume errors, propagating to even greater errors for body fat estimates (19,20). In the mid-1980s, modifications were made so that ADP could be performed in isothermal and adiabatic (no gain or loss of heat) conditions. The results were compared to UWW and, although the comparisons were favorable ($R^2 = 0.93$; SEE = 6.7%) (21), the impracticality of the technique in its present form made its widespread applicability unrealistic.

The breakthrough in more accurate assessments of body volume by ADP came from the work of Dempster and Aitkens (22) and McRory et al. (23), who found reliable and valid estimates of body volume and density could be achieved with their development on new methodology using ADP. The current state-of-the-art system, BOD POD (see figure 3.1), has overcome many of the previous technique issues and demonstrated improved precision and accuracy (24,25).

ADP relies on the physics of Boyle's Law, where pressure (P) and volume (V) are inversely related if temperature is held constant, or

$$\frac{P_1}{P_2} = \frac{V_2}{V_1} \tag{3.8}$$

So the quantity of compressed air will decrease its volume in proportion to the increasing pressure. However, under adiabatic conditions, the temperature of air does not remain constant as its volume changes and the kinetic energy of the molecules changes. The relationship between pressure and volume therefore changes under adiabatic conditions and is described by Poisson's law

$$\frac{P_1}{P_2} = \left(\frac{V_2}{V_1}\right)\gamma \tag{3.9}$$

where γ is the ratio of the specific heat of the gas at constant pressure and at constant volume (26). So for a given gas volume and volume change, the observed change in pressure will be less in isothermal conditions versus adiabatic conditions. Earlier ADP techniques did not account for this difference and hence the significant volume

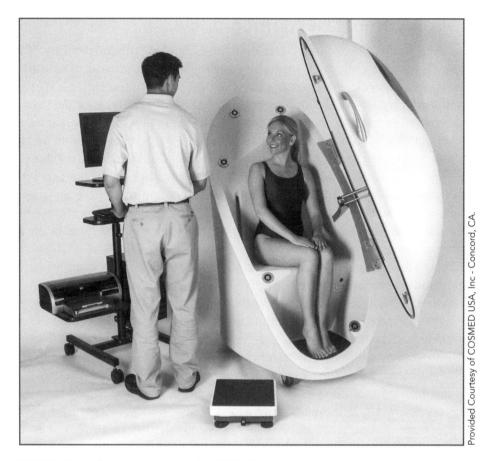

Provided Courtesy of COSMED USA, Inc - Concord, CA.

FIGURE 3.1 Subject sitting in the BOD POD test chamber.

measurement errors that were commonplace. When the BOD POD was developed, it used Poisson's law to determine body volume and thus corrected for this potential volume measurement error.

The BOD POD (Life Measurement Instruments, Inc.), as first described by Dempster and Aitkens (22), consists of a single structure containing two chambers that are separated by a molded fiberglass seat. The subject sits in the test chamber (~450 L [~119 gal]), opposite the reference chamber (~300 L [~79 gal]) (see figure 3.2), and enters through a door that can be sealed by a series of electromagnets. A computer-controlled diaphragm is mounted on the front wall between the chambers and oscillates to create small, sinusoidal volume and pressure perturbations in both chambers that are identical in magnitude but opposite in sign. Fourier coefficients are used to calculate the pressure amplitude at the frequency of oscillations. Because the perturbations are small relative to the ambient pressure of the chamber, Poisson's law states that the ratio of the volumes of the test and reference chambers is equal to the ratio of their pressure amplitudes. The gas composition, and therefore γ in the pressure–volume relationship, is held constant in the two chambers by using an air-circulation system that mixes the air between them. Because any air added to one chamber is subtracted from the other, there is no effect on the equality of the perturbations. In addition, any

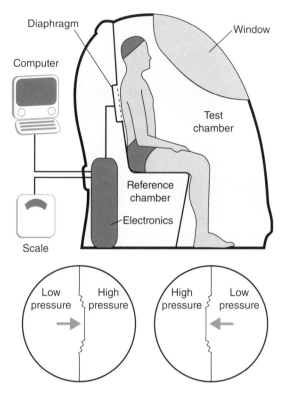

Moving diaphragm produces complementary
pressure changes in the chambers

FIGURE 3.2   Schematic of the internal components of the BOD POD Body Composition
System.

Provided Courtesy of COSMED USA, Inc - Concord, CA.

effect of temperature change during the measurement period is essentially eliminated
with the use of sinusoidal perturbation and Fourier coefficients.

When measuring a person, both the human body and the volume of air within the
lungs will maintain a constant temperature while the person is in the chamber. However, conditions of the test are not adiabatic because of air close to the skin, clothing,
and hair. Consequently, subjects must wear a swimsuit and swim cap during the test
to minimize errors associated with air trapped in clothing and hair. The negative
volume effects of the skin surface area on the total body volume can be determined
by applying a constant (k) to the Dubois formula (1916) to estimate body surface
area (BSA), where

$$\text{BSA (cm}^2) = 71.84 \times \text{Weight (kg)}^{0.425} \times \text{Height (cm)}^{0.725} \qquad (3.10)$$

and

$$\text{Surface area artifact (L)} = k(\text{l/cm}^2) \times \text{BSA} \qquad (3.11)$$

It is also necessary to adjust for the volume of air in the lungs and thorax, called
thoracic gas volume (V_{TG}), by completing a procedure very similar to that used in the

plethysmographic measurement of V_{TG} in pulmonary function testing (27). During the body volume measurement, the subject wears a nose clip and breathes normally through a tube. After two or three normal breaths, the airway is occluded for 3 s at midexhalation. During this time, the subject gently puffs against the occlusion by alternately contracting and relaxing the diaphragm while the airway pressure is measured, yielding a value for the volume of exhaled air that comes into contact with the chamber gas volume. By comparing the degree of change in airway and chamber pressures, calculation of V_{TG} can be completed using the manufacturer's proprietary methods (25).

There are two variables used to ensure that this procedure is being properly performed, the "figure of merit" and airway pressure. The figure of merit (described in detail by Dempster and Aitkens [22]) examines the level of agreement between scaled and translated pressures measured inside the chamber and in the breathing airway, with smaller values indicating closer agreement. Values higher than 1.0 may indicate leaks at the mouth seal, absence of an adequate nose clip, excessive cheek puffing, or contraction of the abdominal muscles (25). Airway pressure is also evaluated to verify that it is not too high ($\geq$35 cm [$\geq$14 in.] water), which may point to closure of the glottis or significant alveolar compression.

Once the uncorrected body volume (V_{UC}), surface area artifact (SAA), and V_{TG} are obtained, the corrected body volume can be calculated as follows:

$$\text{Body volume (L)} = V_{UC} \text{ (L)} - \text{SAA (L)} - 40\% \ V_{TG} \text{ (L)} \tag{3.12}$$

where V_{TG} has been corrected by 40% because of the compression factor seen in isothermal air volumes versus adiabatic volumes. This corrected body volume along with body mass (obtained using a high-precision scale) can be used in equation 3.1 to calculate total body density. Total body density can then be used in an individually appropriate percent body fat equation (i.e., Siri or Brozek) or used as part of a multicomponent model.

Although both the precision and accuracy of the BOD POD for measuring inanimate objects are excellent (22,28), the percent body fat estimates in humans have proved less accurate. Earlier reviews examining the BOD POD's reliability and accuracy (24,25), in addition to more recent studies examining large, heterogeneous samples (29,30), found good to excellent results. Fields et al. (25) summarized SEEs for the BOD POD percent fat versus UWW percent fat, ranging from 2% to 3% in both adults and children. In direct comparison with UWW, the within-day test-retest reliabilities of the BOD POD are similar in some studies but not in others. For example, in one study, both methods show an excellent coefficient of variation of about 2% fat (23), and in another study, both methods show a lower reliability of 4% fat (31).

Regarding the validity of ADP, the consensus of published reviews is that ADP is a valid technique for use in many populations, including children, the elderly, obese subjects, and athletes (32). However, differences among studies reflect a lack of standardized protocols and precise control of technical factors. Comparing BOD POD to multicomponent models is not a valid approach to test the accuracy of the method because the calculated body density is the same value for the two approaches.

In conclusion, the densitometric techniques of UWW and ADP are both useful laboratory methods for the assessment of body composition, but they are limited as two-component models that are based on the assumption that the density of FFM is constant. Using DXA as a criterion variable, Fields et al. (25) and Going (8) showed a general agreement among several studies. In individuals or populations where these assumptions are not met, the preferred use of these techniques would be in combination with other methods and incorporated into multicomponent models. When other such methods are unavailable, it is essential to have an understanding of how different sources of variation affect body composition and then carefully interpret the results.

Total Body Water

Water is the most abundant component in the human body comprising about 60% of body mass in the reference man (33). Because it is mostly found in the fat-free body in a relatively constant amount, assessment of body water has been of interest as a method of body composition assessment for almost 100 years. Unlike the other molecular body components, the water component consists of a single molecular species (H_2O), which simplifies the task of its measurement. Water's characteristic as a singular molecular species offers itself to the use of the dilution principle, which in its simplest form, states that the volume of the component is equal to the amount of isotope added to the component divided by the concentration of the isotope in that component (34,35).

In 1915, the dilution principle was first used in the study of human body composition (36) when the use of a red dye to measure the plasma volume was extrapolated. The investigators verified that the concentration of the dye after mixing was not constant because it "disappeared" from blood plasma. Using a mathematical approach, a reasonable estimate was made to calculate the volume of plasma in which the dye was first diluted. Following this investigation and using the same principle, tracer material was injected intravenously and allowed to reach a uniform distribution, and from the dilution achieved at equilibrium, the constituents of the body were measured. Both radioactive and stable isotopes were thus used to measure the potassium and sodium of the body (35).

Tritiated water was first described by Pace et al. (37) as an isotope for measuring TBW. The main advantage of using tritium (^3H), the radioactive isotope of hydrogen, is that it is readily available and easily assayed by scintillation counting. On the other hand, a large amount of tritiated water must be administered to obtain adequate precision, eliminating its use in cases where the use of radionuclides is restricted (38).

Currently, deuterium (^2H) and oxygen-18 (^{18}O), which correspond to nonradioactive stable isotopes, are the most commonly used isotopes for the measurement of TBW. Oxygen-18 has the advantage that its dilution space more closely approximates TBW, but it can be adequately measured only by isotope ratio mass spectrometry, and the cost of ^{18}O-labeled water is about 15 times more than that of deuterium (39). Thus, deuterium is the most frequently used isotope to estimate TBW because it is a stable isotope and easy to obtain and has lower costs than tritium or oxygen-18 with no radioactivity exposure (39,40). Deuterium can be measured by infrared spectrometry but preferably by mass spectrometry. Greater technical errors have been found using the infrared approach.

When using isotope dilution, particularly deuterated water, two body fluid samples from urine, blood, or saliva are collected: one just before administration of the deuterium dose to determine the natural background levels and the second after allowing enough time for penetration of the isotope (39,40). If the amount of isotope is known and the baseline and equilibration concentrations are measured, the volume in which the isotope has been diluted can be calculated (39).

There are four basic assumptions that are inherent in any isotope dilution technique.

1. *The isotope is distributed only in the exchangeable pool.* None of the commonly used isotopes are distributed only in water. But tracer exchanges with nonaqueous molecules are minimal, and consequently, the volume of distribution or dilution space of the isotope can be determined, albeit slightly greater than the water pool (34). Deuterium exchange with nonaqueous molecules is estimated at 4.2% in human adults (41).

2. *The isotope is equally distributed within the pool.* Isotopic tracers are identical to body water, except for differences in molecular weight, which can lead to isotopic fractionation. Isotopic fractionation corresponds to the process that accounts for the relative abundances of isotopes and consequent redistribution of isotopes within the body (39). Samples collected from plasma, urine, and sweat do not show fractionation, whereas samples from water vapor do (39).

3. *Isotope equilibration is achieved relatively rapidly.* The equilibration time corresponds to the point where all body fluid compartments have the same proportion of the isotope (42). The rate of equilibration as a function of the route of deuterium oxide administration has been investigated by Schloerb et al. (43), who observed that equilibration was reached 2 h after intravenous administration and 3 h after subcutaneous or oral administration. Wong et al. (44) verified that the time to equilibration was approximately 3 h regardless of whether plasma, breath carbon dioxide, breath water, saliva, or urine was sampled. Schoeller et al. (45) detected less TBW at 3 h than at 4 h after oral isotope administration. Considering these findings, equilibration time for TBW was set at 4 h (39,42), with the exception of patients with expanded extracellular water compartments, where 5 h equilibration time is required (46,47). Considering TBW assessment, urine has demonstrated a low isotope enrichment relative to venous plasma water (44). Still, it is important to consider voids after tracer administration. Three voids are recommended after the dose when urine is used as the biological sample (44).

4. *The tracer is not metabolized during the equilibration time.* Body water is in a constant state of flux. In temperate climates, the average fractional turnover rate in adults is 8% to 10% each day (48). This turnover comprises inputs of water from beverages, food, metabolic water produced during the oxidation of fuels, and water exchange with atmospheric moisture. The inputs are balanced by an output of water in the form of urine, sweat, breath water, or transdermal evaporation (49). This constant turnover has led to two approaches when assessing TBW: the plateau method and the back-extrapolation, or slope-intercept, method. For body composition research, the plateau method is the usual approach. Deuterated water is administered, samples are collected for 3 to 5 h, and TBW is calculated from the samples collected before and after the enrichment has reached a plateau, or a constant value (50).

In summary, TBW can be accurately measured; however, there is no easy method. Cost of equipment and technical issues make it a laboratory method that takes considerable skill to measure.

Dependency on Two-Component Model

Because water is the most abundant constituent of the body (2,51,52), TBW can be a useful tool to estimate body composition by considering assumptions that are based on biological, chemical, and physical properties of FFM. The principle behind hydrometric models is that lipids are hydrophobic and thus free of water. Consequently, body water is restricted to the FFM component and maintained in a relatively constant amount in the general population with a 72% to 74% mean hydration level and a standard deviation (SD) of 2% to 3%.

PRACTICAL INSIGHTS

There are many reasons to measure hydration level. While a TBW measurement using isotope dilution will give you a volume (L), that alone will not tell you the hydration level of the person being measured. In order to know hydration, you also need a measurement of fat-free mass because the majority of water resides in this component. The fraction of fat-free mass that is water is a relatively stable amount (72% to 74%), and variations from this can be detected. Hydration levels are critical to assess because overhydration (where water content is greater than 76%) and dehydration (where water content is less than 70%) will threaten the health and welfare of the individual. There is clinical value in measuring this in several disease states as well as during and recovering from surgery.

The calculation of FFM from TBW depends on the assumption of a constant hydration of FFM (39). Pace and Rathburn (53) have reviewed chemical analytical data from several mammal species and observed that the FFM hydration corresponded to 73.2%. Indeed, FFM hydration is extremely stable in healthy mammals, including humans (54,55), which thus establishes the following relationship (equation 3.13):

$$FFM = TBW/0.732 \qquad \textbf{(3.13)}$$

Understanding that body mass equals the sum of FM and FFM, it is possible to derive FM by subtraction (equation 3.14).

$$FM = body\ mass - FFM \qquad \textbf{(3.14)}$$

Given that the hydration level is within 2% to 3% in a healthy population (SD), the error in estimating percent fat from assessing TBW is ±3% fat (SEE). Thus, this laboratory method is not quite accurate enough to be a reference method; however, it is often included as part of the four-component model, which is a reference method.

Precision and Accuracy

The precision of measuring TBW depends on the dose of the isotope that is being administered as well as the analytical method that is chosen (39). Overall, considering the TBW measurement with mass spectrometry, particularly high-precision isotope ratio mass spectrometry, very small excesses of deuterium or oxygen-18 can be detected, with a precision ranging from 1% to 2% (39,41,42,56).

The accuracy of dilution techniques to estimate TBW is excellent, unless there is failure to reach equilibrium, which may be the source of bias in the estimation. The accuracy of the technique is dependent on the estimate of nonaqueous exchange, corresponding to about 1% (39). The major source of error arises when using TBW to estimate FFM, particularly when using two-component hydrometric models, which are dependent on the 73.2% constant for FFM hydration because considerable variation in hydration level occurs from person to person. Thus, using multicomponent models will help increase the accuracy of FFM estimation. If a protocol could be developed for management of total water 24 h before the test, holding the hydration level to 72% to 73% (SD = 1.0), then TBW could become a reference method for estimation of fat and FFM.

Limitations

An important consideration when using two-component hydrometric models is the age at which chemical maturity is achieved, meaning that its determination is required to know when the FFM hydration reaches the adult values. The concept of chemical maturity was first defined by Moulton as "the point at which the concentration of water, proteins, and salts becomes comparatively constant in the fat-free cell," which was "the point of chemical maturity of the cell" (57, p. 80). In the first two decades of life, the relative water content of FFM decreases while body density increases, which needs to be considered when using two-component hydrometric models in pediatric ages. Fomon et al. (58) and Lohman (59) have provided information about FFM hydration and density during growth, and reference data based on a large sample of children and adolescents have been presented by Wells et al. (60) using a four-component model of densitometry, hydrometry, and DXA bone mineral (figure 3.3).

Considering this important aspect of growth, it is important to understand that constants based on adult values should not be used in younger ages where chemical maturity of FFM has not been achieved. Also, a higher hydration level in children affects the accuracy of densitometry-derived body composition.

Assessing body composition has also played an important role in monitoring athletic performance and training regimens (9). Variability in the density and chemical composition of the FFM is the primary factor limiting the accuracy of two-component models, including hydrometric models for body composition estimation (3,61,62). In athletes, variability in FFM density and composition has been reported (63-66). Modlesky et al. (63) verified an increase in FFM hydration in male weight trainers probably caused by an increase in skeletal muscle mass because water comprises about 74% of skeletal muscle. Similar results were reported by Withers et al. (67) for bodybuilders during preparation for competition. Other investigations verified that the

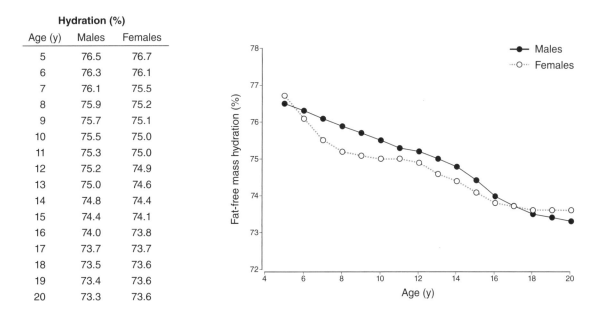

Hydration (%)		
Age (y)	Males	Females
5	76.5	76.7
6	76.3	76.1
7	76.1	75.5
8	75.9	75.2
9	75.7	75.1
10	75.5	75.0
11	75.3	75.0
12	75.2	74.9
13	75.0	74.6
14	74.8	74.4
15	74.4	74.1
16	74.0	73.8
17	73.7	73.7
18	73.5	73.6
19	73.4	73.6
20	73.3	73.6

FIGURE 3.3 FFM hydration for pediatric ages.
Adapted from Wells (2010).

composition and density of FFM did not differ from the established values in athletes (68-70). This lack of agreement among studies limits the usefulness of hydrometric two-component models to estimate body composition in athletic populations.

Use in Four-Component Model

As discussed in chapter 2, at the molecular level, FFM can be partitioned into several molecular components, including water, mineral, and protein (71). Multicomponent models share in common their developments from simultaneous equations, which may include two or more unknown components. As a general rule, for each unknown component, there must be one independent equation that includes the known component (e.g., TBW) and/or the measurable property (e.g., body volume) (72). Even a small change in water can produce a measurable change in body mass; thus, determining TBW is central to accurately estimating body composition (39).

The physical density of the molecular components is of extreme importance for methodological advances. The calculated and assumed constant densities of combined molecular level components are the basis of two, three, and four molecular component level models.

Four-component models control for biological variability in TBW, bone mineral mass, and residual and can be generated using the following concept (73):

$$\frac{1}{D_b} = \frac{FM}{FM_D} + \frac{TBW}{TBW_D} + \frac{Mo}{Mo_D} + \frac{Res}{R_D} \tag{3.15}$$

where D_b is body density, FM is fat mass, TBW is total body water, Mo is bone mineral, Res is residual, and D is density.

By assuming the densities of the molecular components, it is possible to derive each fraction of the equation, considering the density of water at 36 °C (96.8 °F) is 0.9937 g/mL (0.03589969 lb/in.^{3}) and 0.9934 g/mL (0.03588885 lb/in.^{3}) at 37 °C (98.6 °F) (74).

The independent inclusion of TBW measurements and bone mineral in multicomponent models features a major advantage by controlling for much of the intersubject biological variability in FFM density and composition (73). It is important to note, however, that when the water content of the FFM variation is greater than 3% (SD), it is likely that technical error in the measurement of the TBW method is affecting the four-component model by decreasing its accuracy in estimating percent fat (74,75). For example, Clasey and colleagues (74) found a 5% variation in the variability of body water content in FFM. This large amount of variation is likely to be caused by measurement error.

Recommended Protocol

Every aspect of the measurement, including preparation of the participant, dosing, sampling collection, and isotope analysis, needs to be carefully carried out. The most rigorous preparation is required for TBW extrapolation. Schoeller (39) has described one recommended procedure for measuring TBW by the plateau method in adults, which is summarized as follows.

- The participant should be fasting overnight and should not drink any fluids after midnight. The participant should also refrain from exercise after the final meal and avoid excessive insensible water loss attributable to high ambient temperatures.
- Baseline physiological sample (urine, saliva, or plasma) should be collected.
- The body mass of the participant should be measured with minimal-weight clothing.
- A weighed dose of isotope should be administered by mouth. The capped container should be rinsed with 50 mL (1.7 oz) of water and this water administered to the participant.
- The participant should not take anything by mouth during the sample collection period.
- If saliva, plasma, or breath water is sampled, postdose samples should be collected at 3 and 5 h after the dose. If there is excess extracellular water, samples should be collected at 4 and 5 h after the dose.
- If urine is sampled, the subject should void once before the previously mentioned times, and this specimen should be discarded. Two specimens should then be collected at the prescribed times.
- Samples should be stored in airtight vessels until analysis.
- Enrichment of the two postdose samples should agree within 2 SDs of the particular assay.

The method of analysis is dependent on the choice of tracer. For deuterium analysis, mass spectroscopy is necessary (39,40).

In a generic form and under ideal conditions, the calculation of the isotope dilution space (N) involves the application of the dilution principle according to the equation (76)

$$N = \frac{\left(\dfrac{WA}{a}\right)(S_a - S_t)f}{(S_s - S_p)} \qquad \textbf{(3.16)}$$

where N is isotope dilution space in grams, W is the mass of water used to dilute the dose, A is the administrated dose, a is the mass of the dose used in preparing the diluted dose, f is the fractionation factor for the physiological sample relative to body water, S_a is the measured value for the diluted dose, S_t is the value for the tap water used in the dilution, S_s is the value for the enriched physiological sample, and S_p is the value for the predose physiological sample.

The value of the physiological sample can be obtained by the plateau method or by back-extrapolation to the time of the dose (39,42,50), and subsequently equation 3.4 can be applied to the dilution space calculation. The final step in determining TBW from isotope dilution is to correct the previously calculated isotope dilution space for exchange with the nonaqueous components. Deuterium overestimates the body water pool by 4.2% in adults and children. Expressing the water content as a percentage of FFM (where the four-component model is used) is a final indication of the method's accuracy with an expected variation of 2% to 3% (SD).

Total Body Potassium Counting

Total body potassium can be measured from 40K using whole-body counters. The whole-body counter does not expose the participant to any radiation, but using either a liquid scintillation or an NaI crystal as a detector, the whole-body counter counts radioactivity from radionuclides, such as ^{40}K, which emit gamma rays from the body (0.01% of all potassium is naturally radioactive). The supine participant lies in the counter for 6 to 30 min, depending on the detector. Crystal detectors need longer counting times to obtain the same level of precision as shorter times for liquid scintillation counters. The gamma rays detected per minute are directly proportional to the size of the BCM or FFM. To compute BCM in adults, a potassium-to-nitrogen (K-N) ratio is used as follows: BCM (kg) = 0.00833 × K (mmol/kg), assuming a 3 mEq/g K-N ratio and 4% nitrogen based on wet tissue weight. Various equations have been developed for infants and children and to account for extreme over- or undernourishment. Similarly, TBK-FFM ratios have been developed for each population (77). In adults, the potassium content of FFM has been established as 2.66 g/kg for men and 2.50 g/kg for women (78).

Whole-body counter precision for adults is 2% to 5% (77,79). Lower amounts of radioactivity from infants and young children result in poorer precision (77). In the early part of the century, there were approximately 11 whole-body counters in the United States and 30 worldwide (80). Early work validating body potassium as a measure of muscle mass was carried out in domestic animals including sheep, cattle, and swine with dissection and chemical analysis as the reference method (80,81). For humans, the potassium equations are well validated against the reference method of MRI for skeletal muscle (SM) as follows:

1. Ratio model, accounting for 95.9% of the interindividual variation, which performs well in participants aged <70 years, is

$$SM \text{ (kg)} = 0.0085 \times TBK \text{ (mmol)} \tag{3.17}$$

2. Multiple regression equation, accounting for 97% of interindividual variation in MRI-measured SM mass, is

$$SM \text{ (kg)} = (0.0093 \times TBK \text{ [mmol]}) - (1.31 \times sex) + (0.59 \times Black) \\ + (0.024 \times age) - 3.21 \tag{3.18}$$

where sex is 0 for women and 1 for men; Black is 1 for African Americans and 0 for other racial/ethnic groups (82).

Dual-Energy X-Ray Absorptiometry

Another laboratory method of body composition assessment is dual-energy X-ray absorptiometry (DXA). Before the development of DXA (in the 1990s), single-photon absorptiometry (SPA) and dual-photon absorptiometry (DPA) were used to estimate bone density and regional body composition (83). DPA was developed in the 1980s using radionuclides to provide the dual-energy photon needed to determine areal bone mineral density for the diagnosis of osteoporosis. The replacement of the radionuclides with X-ray led to DXA.

Another application came later estimating soft tissue composition, when it was discovered that quantitative estimates of soft tissue could be made from nonbone areas of the body (84). DXA is a fast, relatively noninvasive technique that can measure bone density for the diagnosis of osteoporosis, whole-body composition, and regional composition. However, DXA equipment is expensive, requires a trained technician for both scanning and analyzing the scan, and involves exposing patients to a small dose of radiation.

Briefly, this method utilizes an X-ray beam consisting of two energies that are passed through the human body. The two energies pass through soft tissue and bone at differ-

PRACTICAL INSIGHTS

An advantage of DXA is its ability to measure soft tissue lean mass, both whole body and regional, from a total body scan. This is a feature that is currently underutilized in clinical settings but has great potential to help monitor conditions such as lean mass gain and loss. For example, older individuals tend to lose lean mass and gain fat mass while maintaining their body weight, a condition called sarcopenia. A DXA scan could show these body composition changes as well as exactly where on the body these changes are occurring. With exercise counseling, a person could undertake a targeted exercise program to attempt to stave off or reverse some of these changes.

ent rates, depending on the composition of the tissue. This is done while a person lies supine on the table of the instrument. Although there are manufacturer differences in the manner in which this is done, the general principle is the same for all instruments. Detailed descriptions of this methodology are available elsewhere (85,86).

X-rays pass through human tissue and are attenuated to different degrees based on the composition of the tissue. Bone is a dense material that attenuates the X-ray more so than both lean and fat tissues, whose densities are lower than bone but still different enough to enable these two components to be distinguished from one another. Therefore, DXA is a three-component model of body composition assessment: bone mass, soft tissue lean mass, and FM. In this section, we will explore the use of DXA to estimate each of the three components.

The X-ray beams are introduced from below the table upon which the patient is lying and are detected by an arm above the patient that moves in tandem with the X-ray source (see figure 3.4). Some manufacturers use a pencil-ray beam, and others use a fan beam (a narrow fan beam has also been introduced). The scanning proceeds from head to toe in a rectilinear fashion for pencil beams and a sweeping action for fan beams. It should be noted that the orientation of the X-ray beams relative to the person being measured is such that only a two-dimensional view is possible. This

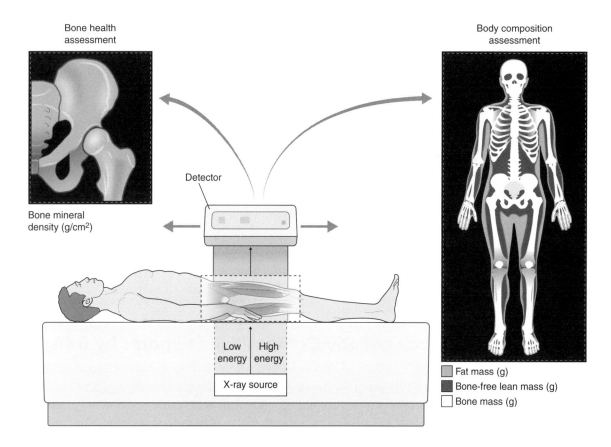

FIGURE 3.4   Schematic of DXA principles.
Adapted from Toombs et al. (2012).

relates to one of DXA's limitations, which is the loss of accuracy when thicker people are measured (also known as beam hardening) (85). Different scanners attempt to adjust for this by providing several scanning modes, with slower scans performed for thicker people (based on the patient's body mass index [BMI]). This feature is automatic or can be manually initiated by the technician on most scanners.

The attenuation of the X-ray that is passed through human tissue in this two-dimensional view is used to determine composition. The areas that are scanned consist of many thousands of pixels that the computer software identifies. Each pixel is first quantified as containing bone or consisting of soft tissue with no bone. Bone-containing pixels are very clearly discerned from nonbone-containing pixels because bone attenuates X-rays to a larger degree than does soft tissue. It is not possible for the DXA software to determine the soft tissue composition of pixels containing bone because the bone's attenuation of the X-ray dominates that pixel.

Because any bone-containing pixel also contains soft tissue, the software estimates the amount and composition of that soft tissue by using data from adjacent pixels that do not contain bone. This approach is a major limitation where up to 40% of the pixels must be estimated for soft tissue composition (83). In nonbone pixels, the attenuation of the X-ray is further examined to distinguish between fat and lean tissue, which attenuate the X-ray to different degrees (both much less than bone tissue).

Lower accuracy of scanning regions with complex bone geometry (such as the trunk) versus areas with more simple bone geometry (such as the appendages) has been described (75). The exact processes that are used to determine composition vary by manufacturer and are not fully available to the scientific community because of the proprietary nature of the software. Early software versions underestimated body fat in the trunk and thigh (87-90). Valentine and colleagues (90) also found an underestimation of truncal fat. However, manufacturer software upgrades continually aim to overcome previously identified software limitations (83). Lohman and Chen (83) have outlined changes in software over time by different manufacturers. It is essential to document the version of software used when employing DXA in a body composition study.

In general, DXA shows excellent precision (coefficients of variation of 1%-3%) and good accuracy (between 2% and 3% fat [SEE] in multicomponent models for the measurement of body composition, which makes it a desired laboratory technique) (9,86). DXA's ability to measure total bone mass allows it to be used in multicomponent models to avoid the assumption that bone mass and density values are constant components of FFM.

Accuracy of Body Composition Estimates by DXA

DXA percent fat can be validated against the four-component model, unlike densitometry, which cannot be directly validated using the four-component model because it provides one of the variables used to estimate fatness. Several research studies have compared DXA percent fat with the four-component percent fat, and SEEs are typically around 3% (75,86). Systematic differences exist among studies between the mean percent fat from DXA (using various hardware and software) versus the four-component model as well as in different populations under study. In general, these differences are between 1% and 2%; however, in highly lean or obese populations, a larger difference is often found (91). Given these limitations, DXA is classified as a laboratory method rather than as a reference method (9).

DXA lean soft tissue and skeletal muscle mass can be validated best with multi-scan computerized tomography (CT) or by MRI. One of the first validation studies to estimate skeletal muscle was conducted in 17 healthy men and 8 men with acquired immunodeficiency syndrome (AIDS). The SEE between methods was 2.1 kg (4.6 lb), or about 4% (92). If we attribute half of the error to the multicomponent model and half to DXA, we have a SEE of 1.5 kg (3.3 lb) $\left(2.1\,\text{kg}\left(4.6\,\text{lb}\right)/\sqrt{2}\right)$ for each method. Valid equations for the estimation of appendicular skeletal muscle mass on larger samples were developed by Kim et al. (93).

The third component by DXA is total bone mineral. Although DXA bone density of the hip and spine are well established for diagnosis of osteoporosis, in body composition studies, it is the total bone mineral mass that is of interest, providing the only predictable method available for the body mineral content (BMC) for the four-component model. Although the precision of estimating BMC is very high, it is difficult to establish the absolute accuracy of DXA estimates of BMC without cadaver analysis. However, it is the relative variation rather than the absolute value in BMC of the FFM that is of primary interest. Using the four-component model, several research studies have shown that water variation more than mineral content of FFM is associated with changes in FFM density (94,95). Thus, research finds the primary source of variation in the four-component model is water variability, with BMC estimates showing a smaller but important source of variation.

DXA can be used to estimate regional body composition, especially abdominal fat. Compared to MRI and CT measures of the L1/L4 region, the DXA fat content showed good agreement (96,97).

Factors Affecting DXA Body Composition Assessment

DXA's ability to detect changes in composition over time has shown promising results with small mean differences between percent fat changes measured by DXA and multicomponent models (86). Houtkooper et al. (98) found DXA to be a sensitive method for assessing small changes in body composition over 1 year in both exercise-training and no exercise–training groups. Tylavsky et al. (99) compared two different DXA systems (fan beam and pencil beam) and found both approaches estimated lean soft tissue changes reasonably well. In several studies, a considerable individual variation in measuring change was found (86). In general, DXA's ability to accurately measure body composition is comparable or better than other two-component laboratory methods (densitometry, TBW) because it is a three-component model.

There are three major manufacturers of DXA instruments: GE Healthcare (Madison, WI); Hologic, Inc. (Bedford, MA); and Cooper Surgical, Inc. (Trumbull, CT). Each manufacturer offers an assortment of instruments and, over time, a variety of software versions and upgrades. There are consistent differences in body composition results between manufacturers, models, and software versions. Coefficients of variation range from 1% to 7% for FM measured by DXAs made by different manufacturers (85-100). Also, systematic differences have been shown between fan-beam and pencil-beam DXAs (100,101). Some of the newest models of DXA have not been thoroughly evaluated for accuracy (86). This presents problems when comparing people who were measured on different instruments or who were measured on the same instrument with different software versions. Large multisite studies have attempted to account for these manufacturer and software differences by performing cross-calibration studies. Comparisons should be performed with caution when instruments or software versions vary. Technicians should strive to be consistent where possible, including reanalyzing older scans with newer software to be able to compare with a follow-up measure.

In addition to these hardware and software factors, DXA results can be affected by a variety of biological and measurement factors. Normal varying hydration levels (1%-3% variation) have been shown to affect percent fat measurements to a limited extent (75). However, on occasion, a 4% or 5% change in the water content of FFM has a greater effect; thus, dehydration may result in systematic changes in body composition estimates, which should be avoided for accurate body composition assessment. The ingestion of large meals can also affect DXA results (102). With a standardized protocol, DXA can be used to measure changes in body composition. Small changes in body fatness (less than 3%) cannot be detected because of the technical error in fat assessment (102).

Body composition measurements from DXA also vary because of patient positioning. The position of the arms (palms flat on the table versus palms against the patient's side) and the position of the legs (together or apart) will cause body composition estimates to vary. Also, whether the person is measured in the prone or supine position has systematic effects (100). Exercise sessions prior to a DXA scan also increase the error of measurement by small amounts (103). Technicians should take care to be trained properly in patient positioning and to position people in a consistent fashion to ensure high precision and accuracy using standard premeasurement protocols. Recommendations regarding the best practices for DXA measurements are discussed in the following sections.

In a review article, Toombs et al. (86) caution that there are large individual differences between DXA percent fat and the four-component model using 95% confidence intervals rather than SEE (95% confidence intervals are roughly two times larger than SEEs). With a SEE of 3% fat, characteristic of most laboratory methods (two out of three subjects within 1 standard error), we expect larger deviations on occasion; thus, DXA, like body water and densitometry, should be treated no differently. Of course, at the upper and lower ends of body fat, errors become larger, and in some cases, a method such as DXA will systematically underestimate or overestimate body fatness (7,91). In any validation study, it is good to take precautions ensuring that all data were accurately collected. In the study by Houtkooper et al. (98), the authors recommend comparing the sum of parts from DXA with scale weight (95). This important quality control measure is rarely reported. In general, a mean difference of 1 kg (2.2 lb) or less is found between scale weight and sum-of-part DXA weight with an SD less than 1 kg (2.2 lb). Prior et al. (95) found a mean of 0.6 kg (1.3 lb) and an SD of 0.5 kg (1.1 lb). Larger SDs than 1.0 may indicate less reliable data from the DXA whole-body analysis.

Another limitation for the practical use of DXA is the size of the scanning area and the low weight capacity of the table, both of which limit who can be measured. Though some manufacturers have begun to increase the area and weight limit of their instruments, researchers have examined alternate scanning procedures to accommodate tall and broad individuals. For tall individuals, two scans can be completed and summed; one scan would be just the head, and the other would include only the body without the head. This would not be effective for very tall people who still exceed the scan area with the head omitted. Another method involves scanning the patient with bent knees, though scanner arm clearance can be an issue for very tall people. Despite the advantage of a lower radiation dose in the bent-knee position compared to the two-scan method, the bent-knee position results in a lower accuracy compared to a whole-body scan (104,105). The two-scan method provides better agreement with a whole-body scan and is the recommended procedure for tall individuals (104,105).

For broad individuals, two scans can also be performed, one for the left side and the other for the right side. These results are summed to obtain total body composition. Studies have shown that summed half-body scans compare well to whole-body scans, making this an acceptable alternative for broad patients (104,105).

Scan analysis is another source of error for measuring body composition using DXA. The software automatically places lines to demarcate the regions of the body. These lines can sometimes be placed in inappropriate locations and cause inaccuracies, especially in regional composition. Trained technicians should follow manufacturers' guidelines and always examine these automatically placed lines, making adjustments where necessary. Hangartner et al. (106) has shown that accuracy is increased when this practice is employed. Using the sum of parts versus scale weight can be used to assess scan analysis errors.

Recommended Protocol for DXA Body Composition Estimates

Many investigators have reiterated the importance of utilizing standardized scanning and analysis procedures when performing body composition estimates from DXA (100,106-108). The following DXA measurement protocol is recommended.

1. Patients are scanned in minimal clothing.
2. Patients are scanned in a fasted state (12 h) and with an empty bladder.
3. Patients are scanned with no prior exercise (~12 h).
4. Patients are scanned in a supine position with hands palms down and not touching the trunk, arms straight, legs straight, ankles strapped, feet in a neutral position, and face up with a neutral chin.
5. Patients should be normally hydrated and not measured in a dehydrated state.
6. DXA machines should be monitored with scanning phantoms to ensure proper operation and to detect any drift that may occur over time.
7. The sum of lean soft tissue, fat, and bone mineral should be compared to the measurement of total body weight by scale.

The mean difference varies with the DXA baseline system with a 0.5 to 1.5 kg (1.1-3.3 lb) difference in favor of scale weight. The SD of the difference is about 1. If the difference is greater than 2 SDs for a given individual, then the DXA scan analysis should be redone.

In summary, DXA is a laboratory method but not a reference method for measuring body composition where the assessment of bone, lean soft tissue, and fat can be assessed. For percent fat, four-component validation studies indicate a SEE of about 3%. Lean soft tissue and total and regional bone mineral can both be assessed accurately. In addition, appendicular skeletal muscle mass and abdominal fat can be estimated. Precision is very good and allows DXA to track body composition changes well. DXA measurements can be affected by hydration levels, food intake, and factors related to the level of training of the technicians performing and analyzing the scans. Critical to the accuracy and precision are standard measurement procedures that must be followed when using DXA to determine body composition.

Ultrasound

The development of ultrasound devices designed specifically for body composition assessment is a relatively recent introduction to the choices available to practitioners interested in estimating body fatness and its associated health risks. Although many think of ultrasound as a tool for prenatal diagnosis and other biomedical applications, evidence of ultrasound's ability to effectively measure body fat has existed for nearly half a century (109,110). A multitude of studies have shown ultrasound to be a precise and accurate method for measuring tissue thicknesses in comparison with more established laboratory techniques (111-117). However, others have found it to be less accurate and no better than commonly practiced field techniques, such as skinfold measurements or bioelectrical impedance analysis, at assessing body fatness (118-121). An extensive review of ultrasound to assess body fat indicates promise as a reliable and accurate method (116).

Ultrasound imaging employs a pulse-echo technique whereby a transducer or scan head, comprising piezoelectric crystals capable of producing high-frequency sound waves, transmits an ultrasound beam through the skin. When the beam comes in contact with different tissues (fat, muscle, bone), it is partially reflected as an echo that the scan head can detect. Because each tissue type has a different density, it also has a unique acoustical impedance, or its own specific resistance to the ultrasound beam passing through it. As a result, when the ultrasound beam crosses the interface between different tissues, it will reflect the beam at varying echo strengths, and the detector can convert these signals to images to provide depth and tissue-type information. Soft tissue interfaces reflect only a small portion of the sound because biological soft tissues differ little in resistance to sound propagation (122).

The ultrasound transducers used for body composition employ two different modes, A mode, or amplitude mode, and B mode, brightness mode. The A mode, used less frequently, is the simpler form of ultrasound imaging. In this mode, the ultrasound wave travels in a narrow pencil beam and can be used to measure depth and, for body composition purposes in particular, the thickness of subcutaneous fat. As the beam reaches the tissue interface between the skin and the subcutaneous fat, some of the ultrasound is reflected back into the probe. The returned wave is recorded as a deviation from a flat line in the form of a spike; the stronger the returned wave, the greater the height of the spike. The height of the spike is called amplitude. As the ultrasound wave continues deeper to meet the interface between the subcutaneous fat and the muscle, some of the wave is again reflected back, and another spike results. Based on the time difference between these two spikes and knowing the speed of the ultrasound, the distance, or thickness, of the subcutaneous fat layer can be determined.

The B mode works similarly to the A mode, where an ultrasound beam is transmitted and the duration of time between interfaces is calculated; however, when the wave is reflected back to the probe, instead of seeing a spike, the strength of the reflected wave is recorded as a bright dot. The brightness of the dot represents the strength of the reflection, such that the brighter the dot, the stronger the wave's reflection. Rather than a single pencil beam, the B-mode transducer emits rapid, successive ultrasound waves that sweep back and forth, resulting in a sequence of contiguous B-mode lines that are synthesized together to provide a two-dimensional image. This sequence of emitted waves and image creation happens so quickly that the image on the screen is essentially shown in real time. Consequently, the technician can not only see the various tissues but also the amount of compression force being applied and the image quality.

Because B-mode ultrasound provides real-time, two-dimensional images, it is superior to A-mode scans for clinical imaging. It also yields a clearer image of the subcutaneous fat layer that is used for assessing body composition. However, measuring the subcutaneous fat thickness does not have the complexities associated with clinical imaging of deeper tissues. In a cadaver study examining subcutaneous fat thickness, the readings from a more complex B-mode ultrasound were not significantly different from those of a commercially available A-mode ultrasound designed specifically for body composition, and both modes had an accuracy of less than 1 mm (0.04 in.) at most sites (123). This was a limited sample of six cadavers, so further comparisons with greater sample sizes will be needed for a true determination of the strength of the A-mode relative to the B-mode ultrasound for measuring body composition.

The procedure for ultrasound is relatively straightforward. The scan head is positioned on the skin at the site of measurement (see figure 3.5). For optimal wave transmission, an ultrasound gel is used between the probe and the skin. This is because, at interfaces of soft tissue and air or bone, the reflection is nearly 100%, making accurate imaging impossible. The scan head is held at 90° to the skin and should be pressed hard enough only to contact the skin, not compress it. If the transducer is not at a right angle, the image will be blurred. The pulse-echo ultrasound beam is turned on, and the scan head is moved minimally (<1 cm [<0.4 in.]) and carefully so as not to break contact with the skin at the site of interest. For any given site, the scan itself takes just a few seconds, and the signals are converted to an image (see figure 3.6), which can be stored for later analysis and interpretation when the technician identifies tissue boundaries and the software measures tissue thickness.

Although ultrasound technique is relatively simple, it has been limited until recently by its lack of uniform guidelines for its use in body composition assessment. Varied

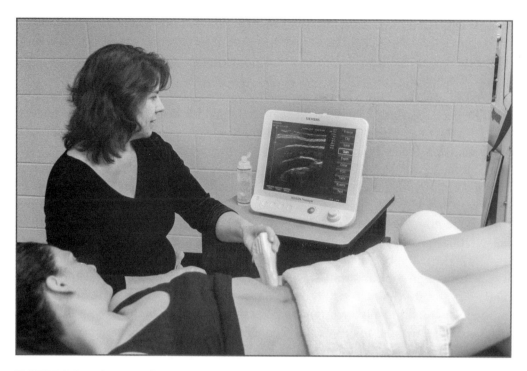

FIGURE 3.5   Ultrasound imaging being performed for subcutaneous adipose thickness.

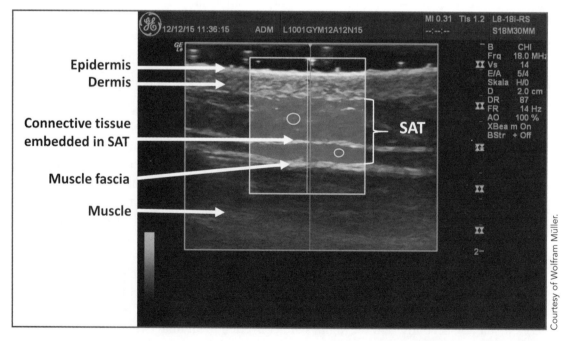

FIGURE 3.6   Example of a B-mode ultrasound image of subcutaneous adipose tissue (SAT).

measurement techniques yield results that are highly dependent on the skill set of the technician (116). For example, varying the force applied to the transducer resulted in up to a 37% change in subcutaneous fat thickness (124). Furthermore, the interpretation is somewhat subjective, and distinguishing tissue interfaces from fascia can be difficult, particularly near the interface of fat and muscle. However, as technicians become more experienced, the interpretation of images improves (109,125,126). The accuracy of measurements can also vary from site to site, so selecting sites that can be consistently measured is crucial to the technique's success.

In an early validation study comparing ultrasound and skinfolds, Fanelli and Kuczmarski (127) found both methods predicted body fatness using the laboratory method of densitometry with a similar degree of accuracy (SEEs = 3.4%-3.8% fat, depending on different combinations of two skinfolds). In more recent results using DXA as a criterion method, Pineau et al. (128) found a 3.0% to 3.2% fat SEE in 83 women and 41 men using A-mode ultrasound at two sites: thigh and postural abdominal wall. In a review of body composition and ultrasound, Wagner (116) concluded that ultrasound can offer a valid approach to body composition assessment.

Protocol recommendations for future ultrasound body composition studies include using the International Society for the Advancement of Kinanthropometry (ISAK) skinfold guidelines, longitudinal scanning of skinfold sites using a high-frequency (12 MHz) linear scanner, and use of generous amounts of gel to ensure minimal compression (124).

Horn and Müller (129) and Müller et al. (130) have set out to improve the consistency of ultrasound by reducing technician error using a semiautomatic image evaluation technique. This novel approach uses newly developed software (130,131) to more accurately detect the contour of the adipose layer. An overall description of the approach is provided by Ackland et al. (9). The technician can visually control what is considered subcutaneous fat by changing a factor, much like a threshold, that determines accepted image variability and then stops where the tissue segmentation matches the adipose tissue contour. The software can then measure a series of distances throughout the region of interest and provide a minimum, maximum, mean, SD, median, mode, and other values. The operator can also decide to select values that include or exclude other embedded tissues, such as fibrous tissues. Using excised pig tissue, the procedure was shown to be highly correlated with vernier caliper measurements (129).

More recently, the technique was used in athletes and found to be more reliable and accurate than the traditional field technique of skinfold thicknesses (130,131). The interobserver error ranged from 0.4 to 0.8 mm (0.02-0.03 in.) among three ultrasound technicians across eight ISAK skinfold sites (131). Additional research by Müller et al. (132) has led to a set of recommendations for ultrasound fat location better suited to the method and a protocol that further minimizes interobserver error, overcoming a number of limitations summarized by Wagner (116). Störchle et al. (133) has applied these standardized sets of sites in normal-weight, overweight, and obese persons and found high reliability. Specialized training in this new approach is offered by International Association of Sciences in Medicine and Sport (www.iasms.org).

PRACTICAL INSIGHTS

Ultrasound has much promise due to innovations in the software systems used for the analysis of the images captured during the measurement for body composition analysis. This method still needs to be validated against a four-component model. If this proves to be valid, it could be both a laboratory method and a field method of body composition assessment since it has promise to be highly accurate and is also portable, unlike all other laboratory methods. It is rare that a highly accurate method is portable and lower in cost. Scientists are now conducting studies to test ultrasound for validity in various populations and under different conditions to make sure it meets the various validity criteria that have been previously mentioned.

These findings are encouraging for the possibility of ultrasound as a viable tool for body composition assessment. As a laboratory technique, it has the advantage over other effective techniques, such as DXA or CT, in that there is no ionizing radiation involved. Its cost, portability, and speed of measurement are also assets compared with other medical techniques. Future work in both reliability and validation studies will likely result in the more widespread use of ultrasound in assessing body composition in both athletic and nonathletic populations. Müller (112) has proposed this approach in the athletic population.

Summary

This chapter reviewed the major laboratory body composition methods: densitometry (UWW, ADP), TBW, TBK, DXA, and ultrasound. These level 2 methods are the next most accurate approaches to whole-body composition assessment after those presented in chapter 2. Although they lack the highest accuracy associated with the level 1 methods, each of these methods is more accessible and cost friendly to those interested in an accurate assessment of their body composition. The chapter provided the theoretical basis, advantages, limitations, accuracy, and precision for all of these laboratory methods. As a result, you now have a working knowledge of how each method's particular sources of variation will affect its ability to effectively measure body composition, which in turn will enable you to provide a more educated interpretation of those results.

4

Body Composition Field Methods

Leslie Jerome Brandon, PhD, FACSM
Laurie A. Milliken, PhD, FACSM
Robert M. Blew, MS
Timothy G. Lohman, PhD

LEARNING OBJECTIVES

After completing this chapter, you will be able to do the following:

- Describe the use of field methods in predicting body composition

- Become familiar with standardized measurement procedures for different field methods used to estimate body composition and obesity

- Compare different field methods used to estimate body composition for their advantages and limitations

- Explain the generalizability of different field methods for estimating body composition

- Describe the advantages of field methods compared to various indexes of weight and height for estimating body composition

Several field methods are used to estimate body composition. In this chapter, we cover skinfolds, circumferences, bioelectrical impedance analysis, and various indexes of weight and height as the major field methods for body composition assessment.

The validity of different body composition field methods and the predictive accuracy of many published equations need to be carefully evaluated. Table 4.1 features subjective ratings of percent body fat based on standard error of estimate (SEE), using validation studies that provide for levels of accuracy for different body composition methods. Under the best conditions, the SEE of percent fat for selected field methods is within 3% to 4%.

Table 4.1 Subjective Ratings of Percent Body Fat (%BF) Based on SEE

SEE %BF	Subjective rating
2.0	Ideal
2.5	Excellent
3.0	Very good
3.5	Good
4.0	Fairly good
4.5	Fair
5.0	Poor

Reprinted by permission from T.G. Lohman, *Advances in Human Body Composition* (Champaign, IL: Human Kinetics, 1992).

PRACTICAL INSIGHTS

SEEs for the estimate of percent body fat are given in table 4.1. Any new body composition field technique should show at least good levels of error for use in the field. No doubt there will be new techniques developed and marketed to the consumer. Practitioners should always ask companies that sell new devices about their error rates and whether their new devices were compared to a four-component model of body composition. Companies should be able to provide you with the validation studies, which should report the SEEs for the comparison to the four-component model. The SEEs should be 3.5% or smaller, otherwise the product may not be worth your investment. Shop around for a method and a device that can meet this accuracy standard.

Skinfolds

Skinfold thickness measurement is one of the longest-standing and widely used approaches for estimating body fatness. This approach is based on the assumption that a fold of skin and the double layer of subcutaneous fat included in that fold are representative of overall body fatness. Because over 50% to 70% of body fat is located subcutaneously, this assumption has been shown to be valid using a combination of selected skinfold measurements (2). In general, several skinfold thicknesses can be used to estimate total percent body fat with a SEE in the range of 3% to 4% (good to fairly good). The skinfold method also allows one to characterize the distribution of the subcutaneous adiposity or fat patterning, which has been shown to relate to obesity and chronic disease risk.

Skinfolds have been found to be one of the most commonly used body composition field methods by practitioners, scientists, and medical and health professionals whether included in percent body fat formulas or reported as independent values (3). The most likely reason is that skinfold calipers are mechanically simple, highly portable, and relatively inexpensive (4). Consequently, once properly trained, clinicians can use them to help monitor health status; athletic trainers can use them to help examine body fatness and physical fitness and related changes; and researchers can use them to survey body composition in large-scale clinical trials and epidemiological investigations.

In 1988, the *Anthropometric Standardization Reference Manual* (5) was published to provide a guide for clinicians and researchers with a standardized set of anthropometric dimensions. Prior to 1988, many definitions of skinfold sites were prevalent in the literature with no standardized protocol. The reference manual was widely used in the literature from 1990 to 2005 and is still in use today by researchers. The International Society for the Advancement of Kinanthropometry (ISAK), established in 1986, published its *International Standards for Anthropometric Assessment* in 2001 on a wide variety of skinfolds, circumferences, and skeletal dimensions. A carefully designed training and certification process was established for anthropometrists. The ISAK approach places emphasis on the raw skinfold data over conversion to body density and percent fat. Both raw skinfold data and skinfolds converted to body density and percent fat are currently being used throughout the world (3).

Many practitioners use fat percentage values for discussion with their clients. Although the relationship of skinfolds to body fatness is well established (2), a multitude of equations currently exist for predicting body density and percent fat, and many of these equations have not been cross-validated or have been found to be less effective in different populations. The majority of these equations are linear equations developed in homogeneous populations that are unsuitable for use in other populations (i.e., a skinfold equation developed in active college males will be less accurate for sedentary postmenopausal women). An exception to this limitation is the Durnin and Womersley (6) equation based on the log of the sum of four skinfolds (biceps, triceps, subscapular, and suprailiac) to estimate body fatness using the two-component model (body density) as the criterion method in four different age groups of adult men and women (6).

The exact measurement protocol used by the investigator to develop the equation is often overlooked by the practitioner. In 1996, Heyward and Stolarczyk (7) published a body composition assessment manual to aid the practitioner in selecting an appropriate equation and measurement protocol. We have simplified the choices and included more recent publications in this chapter and chapter 7.

A more practical approach for skinfolds would be to use a single set of sites that is applicable to everyone, accounting for gender, age, ethnicity, and athletic status. Jackson and Pollock (8) developed skinfold equations for young and middle-aged adult men and women using the sum of three, four, or seven skinfolds coupled with age that provide an estimate of body density and percent fat. Jackson and Pollock found a curvilinear relationship between the sum of selected skinfolds and body fatness. These equations were developed on a heterogeneous sample of adult subjects ranging in age from 18 to 55 years and in the college athletic population (9-11), and they have been cross-validated by other investigators for use in the adult population.

A study by Peterson et al. (12) concluded that both approaches by Jackson and Pollock (8) and Durnin and Womersley (6) underestimate body fat. However, Peterson et al. performed a biased Bland–Altman analysis, as explained by Evans et al. (13), and did not follow the Jackson–Pollock measurement protocol when measuring their skinfolds. One of the best designed studies to develop generalized skinfold equations (sum of three and seven sites) for the athletic population was carried out by Evans et al. (13) using the four-component model as a reference method (see chapter 7).

The equations developed in adults are not suitable for children given their difference in body fatness, fat-free mass (FFM), and fat patterning. A generalized skinfold equation in adult males was developed by Lohman (2) where the actual data from multiple investigators ranging from athletic to obese were combined into one heterogeneous sample to increase the external validity of the equation. Following a similar protocol with adolescent male wrestlers, Thorland et al. (14) determined that an equation with three skinfold sites (triceps, subscapular, and abdomen) was an effective equation, and this equation has become widely used throughout the country (chapter 6) with adolescents. Slaughter et al. (15) developed equations based on skinfolds using a four-component model for estimating percent fat in children.

Skinfold measurement sites described in the following sections will be used for site location based on the sites defined by Jackson and Pollock (8) and by the *Anthropometric Standardized Reference Manual* (5). Also, measurement technique will be discussed.

Other generalized equations for both children and adults have been developed from the research of Stevens et al. (16,17), Jackson et al. (18), O'Conner et al. (19), and Davidson et al. (20). Stevens et al.'s unique approach using the least absolute shrinkage selection operator has excellent external validity where many equations fall short because they do not generalize to other samples. The Stevens approach used a very large national probability sample (National Health and Nutrition Examination Survey [NHANES]) of different ethnicities of children and adults with excellent cross-validation results, using one skinfold and one circumference, with adjustment for menarche, age, race/ethnicity, curvilinearity, and many selected interactions. The Stevens et al. equations were validated with dual-energy X-ray absorptiometry (DXA; Hologic 4500 fan beam) serving as the criterion method. The equations yielded SEEs between 2.6% and 2.9% fat. Because of the limitation in the Hologic model of DXA as a criterion, these equations need to be cross-validated using the four-component model.

Jackson et al. (18), O'Conner et al. (19), and Davidson et al. (20) cross-validated the Jackson–Pollock equations (10) and Durnin–Womersley equations (6) in difference races and ethnicities using DXA as the criterion method. New equations have been developed by Jackson et al. (18) that adjust for some biases found in the original equations. The modified Jackson–Pollock and Durnin–Womersley equations need to be cross-validated in different populations using the four-component models because there may be some bias in the DXA 4500 fan beam, depending on the subject's size, age, and composition. Schoeller et al. (21) report that the DXA 4500 fan beam underestimates fat and overestimates FFM by 5%, thus making generalizability from equations validated with the DXA 4500 questionable in very lean and obese adults and children. Recommended skinfold equations are presented in chapter 7 for children and youth, adults (nonathletic), male adolescent athletes, adult athletes, female athletes, and the elderly.

Skinfold Measurement Preparation

Field assessments of body composition are effective methods for the practitioner to assess body composition, and it is important to consider the participant and recognize the potentially uncomfortable nature from her perspective, especially for children. Before performing any measurements, it is important to keep in mind that the participant may be apprehensive of the information to be found. As such, remember that each participant being measured has the right to confidentiality. Every effort should be made to keep observations and data recording as objective and nonjudgmental as possible. It is important not to react to any measure; simply observe and record. Technicians should be pleasant and respectful to each individual and make the experience a positive one. It is also essential to think of the safety of the subject. Technicians should be careful in the use of pens and pencils while taking measurements and should remove rings, bracelets, or other jewelry that could pose a hazard.

Prior to measuring, verify that your caliper is properly calibrated. At rest, when the jaws are closed, the caliper should read 0 mm (0 in.). Calibration blocks are available that enable you to verify the caliper's ability to accurately read progressive increments (i.e., 10, 20, 30, and 40 mm [0.4, 0.8, 1.2, and 1.6 in.]) and should be used prior to any measurements or if the calipers are accidentally dropped during measurements. Measurers should introduce themselves to the participant and carefully explain all procedures. If the participant seems anxious or is a child, it may be helpful to demonstrate the caliper on the hand of the technician and then on the hand of the subject by measuring the musculature between the thumb and index finger (see figure 4.1).

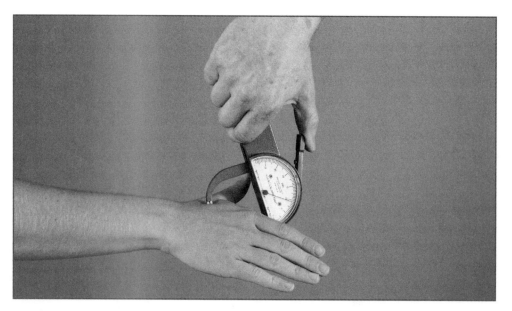

FIGURE 4.1   Demonstration of caliper use on the hand for apprehensive participants.

Skinfold Measurement Techniques and Sites

A right-handed technique that is applicable to all types of skinfold calipers is presented, and this technique is used for each of the sites outlined subsequently. For standardization purposes, skinfolds are typically measured on the right side of the body. Only if a participant has physical injuries or deformities such that the right side cannot be measured should the left side serve as a substitute. Skinfold measurements are recorded to the nearest millimeter (0.04 inch) and should be performed three times at each site with the average of the three measures used as a criterion. If the three skinfolds do not agree within 10% of the mean, repeat all three measurements. Further, a well-trained researcher should be within 10% of an expert for each skinfold site (mean of three measurements) (5).

The skinfold to be measured is formed by using the thumb and index finger of the left hand to grasp and elevate a double fold of skin and subcutaneous adipose tissue approximately 1 cm (0.4 in.) from the location at which the skinfold is to be measured (the direction will depend on the skinfold site). Grasping 1 cm (0.4 in.) away from the site ensures that, when the caliper in the right hand is later placed on the site, it is 1 cm (0.4 in.) away from the fingers grasping the fold. Maintaining the space between the "pinch" and the measurement site is vital because it prevents the pressure of the pinch itself from affecting the natural skinfold thickness. The skinfold is formed by placing the thumb and index finger on the skin, roughly 8 cm (3.1 in.) apart, forming a vertical line along the axis of the future skinfold, and then drawing them toward each other. The 8 cm (3.1 in.) distance may vary based on the size of the subject, with thicker subcutaneous adipose tissue layers requiring a greater separation between the thumb and finger prior to drawing them together. For larger folds, the index finger alone may not be adequate for acquiring the entirety of the skinfold. In this case, the middle finger and, for even larger folds, the ring finger also can be used

PRACTICAL INSIGHTS

Protocols presented here are detailed and should be followed exactly. This is important to ensure the lowest error possible for the measurements that you make. Any errors that occur in each step of the measurement process will accumulate and contribute to the overall error. For example, the procedures for the measurement of skinfold thicknesses first involve locating the measurement site. Measuring the incorrect site can introduce error. Once you locate the site, you will need to grasp the fold properly. There is a technique for grasping the fold and for how and where to place the caliper. Finally, the technician must read the caliper properly. There are other sources of error involved with selecting the correct equation for the person you are measuring, purchasing a high-quality caliper, and performing the calculations correctly. With training, all of these sources of error can be minimized.

to grasp the fold. The skinfold should remain firmly grasped until completion of the measurement (5).

The common principle is that the long axis of the skinfold runs parallel to the natural cleavage lines of the skin at the site to be measured. The raised skinfold should have parallel sides and consist only of skin and subcutaneous adipose tissue (22). A common error in less experienced technicians is to include muscle in the fold, particularly in leaner subjects, at the appendicular regions. If uncertainty exists, the subject can be asked to flex the muscle while the technician holds the skinfold to help determine whether muscle is being included. When using this technique, the skinfold should be released, and the standard measurement procedure reinitiated prior to obtaining the measured value. All measurements should be made while the person's muscles are relaxed. The specific skinfold descriptions and measurement techniques for the triceps, chest, subscapula, abdomen, anterior suprailiac, thigh, and calf are detailed in the following sections.

Triceps

For the triceps, the subject is measured standing, with the arm hanging freely at her side. The measurer stands behind the subject and places the palm of his left hand on the subject's arm above the mark, with the thumb and fingers directed inferiorly. This is a vertical measurement taken at a point midway on the posterior aspect between the lateral projection of the acromion process and the inferior margin of the olecranon process of the ulna. Using a tape measure, measure the distance between these two points along the lateral portion of the arm with the elbow flexed to 90° (see figure 4.2). The tape is placed with its zero mark on the acromion and stretched along the upper arm, extending below the elbow. Mark the midpoint on the lateral side of the arm. Have the subject straighten and relax the arm at her side, and place a second mark, level with the first, on the midline of the posterior arm. The mark should be at the crest of the skinfold, and the measurement should occur midway between the crest and base of the skinfold (see figure 4.3) (5,7).

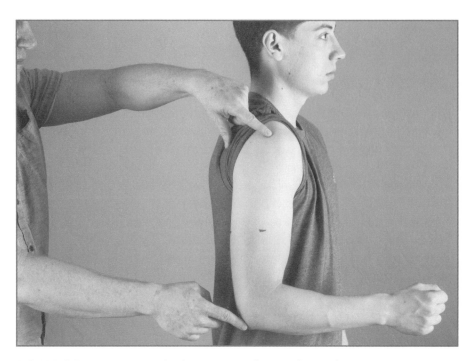

FIGURE 4.2   Positioning the bent arm to locate the midpoint.

Using a professional-grade skinfold calipers, pressure is applied with the thumb to open the caliper jaws, and the opening is slipped over the skinfold roughly midway between the crest and base of the skinfold (see figure 4.4). The calipers are placed perpendicular to the long axis of the fold as pressure exerted by the thumb on the caliper is gradually released, allowing the jaws of the caliper to close on the fold until

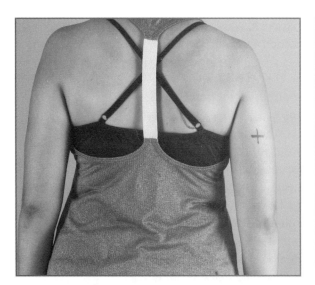

FIGURE 4.3   Point where triceps skinfold measurement occurs.

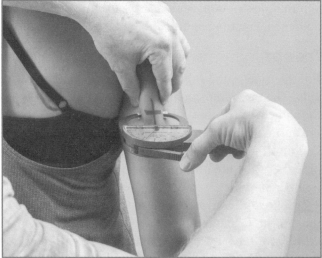

FIGURE 4.4   Grasp and caliper placement for triceps measurement.

the thumb is no longer exerting any pressure. The measured value should be taken between 2 and 4 s after the caliper thumb releases pressure. If the caliper applies force for longer, fluids will begin to exit the tissues within the fold, and the measured value will decrease, resulting in inaccurate measurements. After taking the measurement, the caliper jaws should be opened and removed followed by the release of the skinfold by the left hand. Failure to perform this final procedure may result in a bruising or lacerating pinch to the participant (5).

Chest

The pectoral (chest) skinfold is an oblique fold measured between the anterior axillary fold and the nipple (24,25) along the natural cleavage line. For males, a mark is placed at the midpoint between these landmarks, and for females, at one-third of the distance from the anterior axillary fold (see figure 4.5). With the subject standing and arms relaxed at the sides, the skinfold is picked up 1 cm (0.4 in.) superolateral to the mark while the calipers measure at the mark (see figure 4.6). The measurement is made to the nearest 0.1 cm (0.04 in.).

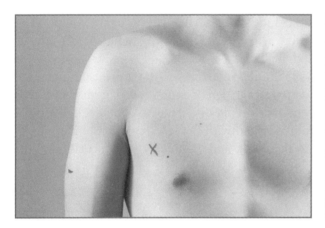

FIGURE 4.5　Chest skinfold site.　　　　　　　　**FIGURE 4.6**　Chest skinfold measurement.

Subscapula

The subscapular skinfold is located just below the inferior angle of the scapula. It falls on a diagonal line running inferolaterally, approximately 45° to the horizontal plane and falling along the natural cleavage line of the skin. The subject is measured standing with the arms at the sides. To locate the site, the measurer palpates the scapula, moving the fingers inferiorly and laterally along the medial border until the inferior angle is identified. If there is difficulty finding the inferior angle because of excessive soft tissue, the subject's arm may be gently positioned behind the lower back to make the scapula more pronounced. Place a mark at the location of natural cleavage directly below the inferior angle of the scapula; this will be the caliper measurement site (see figure 4.7) (5).

The thumb and index finger raise the skinfold approximately 1 cm (0.4 in.) superomedial to the mark, and the thickness is recorded to the nearest 0.1 cm (0.04 in.) (see figure 4.8).

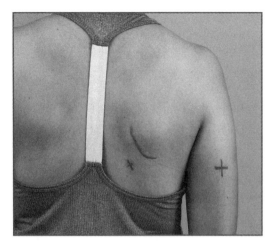

FIGURE 4.7 Subscapular landmark.

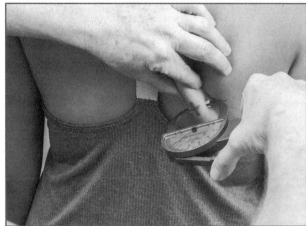

FIGURE 4.8 Subscapular skinfold measurement.

Abdomen

The abdominal skinfold site is marked at 3 cm (1.2 in.) lateral to the midpoint of the umbilicus and 1 cm (0.4 in.) inferior to it (see figure 4.9). The subject relaxes the abdominal wall musculature during the measurement and breathes normally. If movement associated with normal respiration interferes with the measurement, the subject may be asked to hold his breath near the end of expiration. The subject stands with normal posture, and the body weight is evenly distributed on both feet. A vertical skinfold (11) or a horizontal skinfold is raised with the left thumb and index finger positioned 1 cm (0.4 in.) superior to the mark, and the thickness is measured to the nearest 0.1 cm (0.04 in.) (see figure 4.10*a* and 4.10*b*).

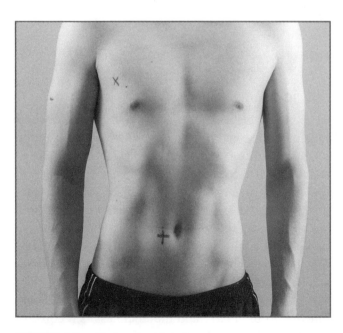

FIGURE 4.9 Abdominal landmark.

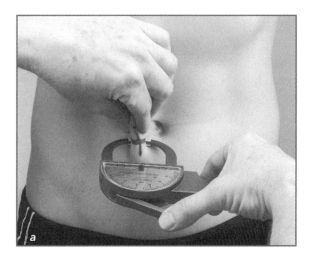 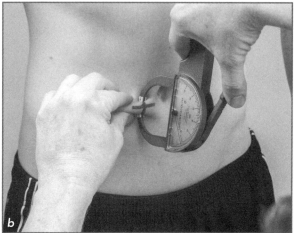

FIGURE 4.10 Abdominal skinfold measurements: (*a*) vertical and (*b*) horizontal.

Anterior Suprailiac

The suprailiac skinfold is an oblique fold located anterosuperior to the iliac crest and posterosuperior to the anterior superior iliac spine originally defined by Jackson and Pollock (8) (see figure 4.11).

With the subject standing with normal posture, arms at sides and feet together, palpate for the iliac crest at the midaxillary line, and move anteroinferiorly to find the anterior superior iliac spine. If necessary, the arm can be abducted slightly to improve access to the site. Place a mark superior to the midpoint of the distance between the bony landmarks following the natural cleavage lines of the skin. This fold runs approximately 45° anteroinferiorly, and the mark will be near parallel or slightly superior to the height of the iliac crest. The oblique skinfold is grasped 1 cm (0.4 in.) posterior to the mark, the calipers are placed at the mark, and the thickness is recorded to the nearest 0.1 cm (0.04 in.) (see figure 4.12) (26).

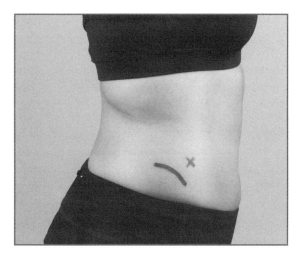 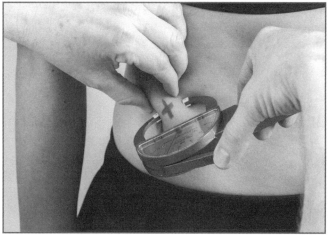

FIGURE 4.11 Anterior suprailiac landmark.

FIGURE 4.12 Anterior suprailiac measurement.

Thigh

The thigh skinfold site is located on the midline of the anterior aspect of the thigh at the midpoint between the inguinal crease and the proximal border of the patella (see figure 4.13). The inguinal crease is located while the subject flexes the hip, and the leg is then straightened to find the proximal border of the patella with the subject's leg at full extension. The distance between these two points is measured and a mark placed at the midpoint.

The vertical skinfold is measured with the subject standing and is aligned with the long axis of the thigh. The body weight is shifted entirely to the left foot while the right foot is positioned slightly forward, flat on the floor, with the knee slightly flexed and the leg relaxed. Be prepared to provide the subject with something for support if maintaining balance is an issue. The thumb and index finger should elevate the skinfold about 1 cm (0.4 in.) proximal to the mark while the calipers take the measurement at the mark to the nearest 0.1 cm (0.04 in.) (see figure 4.14). In individuals with large thigh fat folds, additional fingers may be needed to successfully complete the measurement (5).

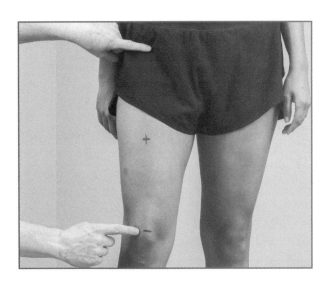

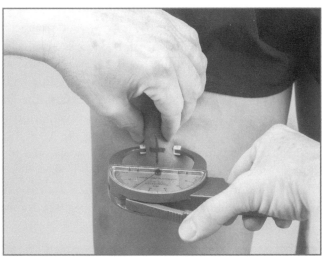

FIGURE 4.13 Thigh skinfold site.

FIGURE 4.14 Thigh skinfold measurement.

Calf

The calf skinfold site is located on the medial aspect of the calf at the level of maximum circumference. The subject should be either seated with the right leg flexed to about 90° with the foot flat on the floor or standing with the right foot on a platform or box so that the knee and hip are flexed to about 90°. Maximum calf circumference is determined by positioning an inelastic tape measure horizontally around the proximal calf and moving it to the distal calf and then back again to determine the level of greatest girth. A mark is placed at this level on the medial aspect of the calf (see figure 4.15). From a position in front of the subject, the measurer raises a skinfold parallel to the long axis of the calf approximately 1 cm (0.4 in.) proximal to the mark

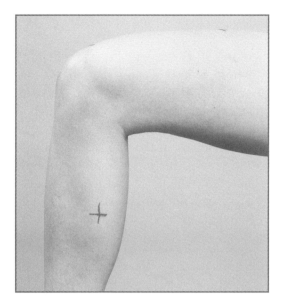

FIGURE 4.15 Calf skinfold site.

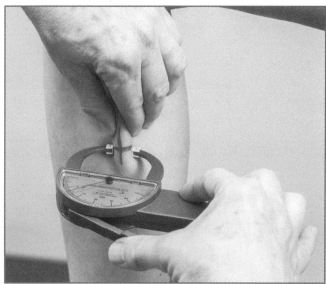

FIGURE 4.16 Calf skinfold measurement.

(see figure 4.16). The thickness of the fold is measured at the mark to the nearest 0.1 cm (0.04 in.) (5,7).

Circumferences

Body circumferences are girth measurements taken at specific body sites that provide information about body composition and nutritional status. Circumferences are measurement friendly, associated with cardiometabolic conditions, and relate to body weight and fat deposits. More than 17 sites for circumference measurements have been used in equations for predicting body fatness in the past several decades (27).

Although many studies have used circumferences for estimating skeletal muscle mass and fat distribution, reproducibility and interrater reliability are challenges for researchers and practitioners. Reproducibility can be increased by giving special attention to positioning the subject, using anatomic landmarks to locate measuring sites, taking readings in millimeters with the tape measure directly in contact with the subject's skin without compression, and keeping the tape at 90° to the long axis of the region of the body under the measured circumference. A well-trained researcher should be, for routine measurements of circumferences, within 2% of an expert (5,28). The circumference measurements should be taken on the right side of the body and, with the investigator taking three trials, should also be within 3% from highest to lowest. If not, the three trials should be repeated.

Circumference Measurement Techniques and Sites

Positioning of the tape for each specific circumference is important for an accurate measure. For each circumference, place the plane of the tape around the site perpendicular to the long axis of that part of the body. For those circumferences typically measured with the subject erect (waist and hip), the plane of the tape is also parallel to the floor.

The tension applied to the tape by the measurer affects the validity and reliability of the measurements. The Gulick II tape is recommended because it applies a consistent amount of tension: 0.11 kg (0.24 lb) each time. If you do not have a tape with a tension device, hold the tape snugly around the body part but not tight enough to compress the subcutaneous adipose tissue. For the arm circumference, there may be gaps between the tape and the skin in some individuals. If the gap is large, a note should be made on the data form, but in most instances, this gap is small and of little concern. Attempting to reduce the gap by increasing the tension of the tape is not recommended. The following description of upper arm, forearm, waist, hip, thigh, and calf circumferences is taken from appendix B in the *Anthropometric Standardization Reference Manual* (5).

Upper Arm Circumference

To measure upper arm circumference, extend the right arm of the participant so that it is hanging loosely by his side with the palm facing inward. At the previously measured midpoint mark (see figure 4.3), wrap the measuring tape around the arm. The zero end of the tape is in your right hand. Once the tape is around the appendage, switch hands so that the zero end of the tape is in your left hand and the other end of the tape is in your right hand, or cross your hands so that the tape overlaps. Pull the tape slightly with your left hand until the appropriate tension is achieved. Hold the tape in place, and record the measurement to the nearest 0.1 cm (0.04 in.) (see figure 4.17).

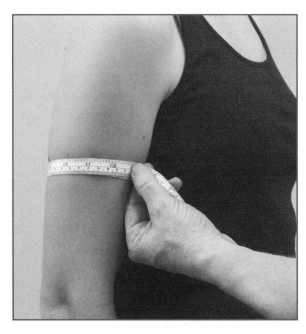

FIGURE 4.17 Upper arm circumference measurement.

Forearm Circumference

To measure forearm circumference, position yourself perpendicular to the long axis of the forearm as the subject has her arms hanging down away from the trunk and forearm supinated (see figure 4.18). Apply the tape snugly around the maximum girth of the proximal part of the forearm. The zero end of the tape is in your right hand. Once the tape is around the appendage, switch hands so that the zero end of the tape is in your left hand and the other end of the tape is in your right hand, or cross your hands so that the tape overlaps. Pull the tape slightly with your left hand until the appropriate tension is achieved. Hold the tape in place, and record the measurement to the nearest 0.1 cm (0.04 in.) (see figure 4.19).

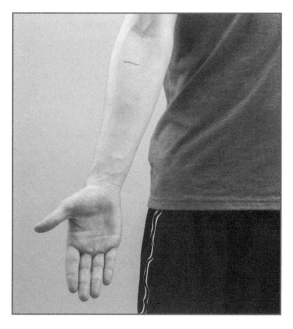

FIGURE 4.18   Forearm circumference site.   **FIGURE 4.19**   Forearm circumference measurement.

Waist Circumference

Waist circumference (WC) and waist-to-hip ratio have been shown to be associated with metabolic and cardiovascular diseases. Correlations between abdominal visceral fat measured by abdominal circumference or WC and cardiovascular risk factors are significant. Abdominal circumference presented similar sensitivity and specificity to ultrasound-measured abdominal visceral fat for identifying the presence of a cluster of at least three cardiovascular risk factors (29). Most of the studies have been completed with adults, but WC has also been shown to provide a simple yet effective measure of trunk adiposity in children and adolescents. This is important because children and adolescents are constantly changing (30).

The standardized description of waist and abdomen circumferences has not been used. Wang et al. (31) completed a study to assess which measurement protocol of four commonly used anatomic sites in this area of the body (immediately below the lowest rib, at the narrowest waist, midpoint between the lowest rib and the iliac crest, and immediately above the iliac crest) yielded a higher reliability. For all four sites, high reliabilities were found, and all sites were similarly associated with total body fat and trunk fat (DXA). Based on the results of the Wang study and because of the ease of measuring minimum WC, we have used that site for all waist measurements. Because there are systematic differences among the four sites in the absolute value, it is essential to receive training in the exact measurement protocol to be effective.

Elliott (32) developed a computer-based tutorial for teaching WC and found good agreement between a well-trained expert and trained participants. Carranza et al. (33) was not successful in following a different protocol for obtaining valid self-measured WC.

WC measurement is easier to make with the participant's shirt removed. If the shirt is not removed, it can be lifted by another technician so you can make this measurement underneath the shirt and against the skin. Women may be willing to tuck their shirts underneath their bra straps. In addition, the participant may need to lower the waistband of his underclothing. Ask participant to stand erect, with feet together and abdomen relaxed. Stand behind participant, and locate the narrowest part of the torso. Stand in front of the participant when taking the measurement. Ask the participant to lift his arms while you place the measuring tape around the narrowest part of the torso. Hold the zero end of the tape in your right hand and the rest of the tape in your left hand. Once the tape is around the torso, ask the participant to relax his arms at his sides. Be sure the tape is in a horizontal plane, evenly placed around the body and not catching on any clothing. It may be helpful to have a second technician check to make sure the tape is horizontal. Once set, switch the zero end of the tape to your left hand and the rest of the tape to your right hand, or cross hands to overlap the tape. Pull the tape lightly with your left hand until the appropriate tension is achieved. Hold the tape in place with your right hand. Record the measurement to the nearest 0.1 cm (0.04 in.) (see figure 4.20) (5).

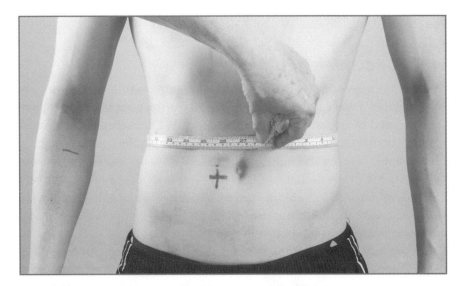

FIGURE 4.20 Waist circumference measurement.

Hip (Buttocks) Circumference

For this measurement, the participant should ideally wear only nonrestrictive underwear (i.e., no control-top panty hose, girdles, spandex, etc.) or a light smock over the underwear. Ask the participant to stand erect with her arms by her sides and feet together. Weight should be evenly distributed on both feet. Squat or kneel down to the right side of the subject. Locate the level of maximum extension of the buttocks. Holding the zero end of the tape in your right hand, extend the measuring tape around the buttocks in a horizontal plane at this level. Once set, switch the zero end of the tape to your left hand and the rest of the tape to your right hand. Pull the tape lightly with your left hand until the appropriate tension is achieved (see figure 4.21). Hold the tape in place with your right hand. Record the measurement to the nearest 0.1 cm (0.04 in.).

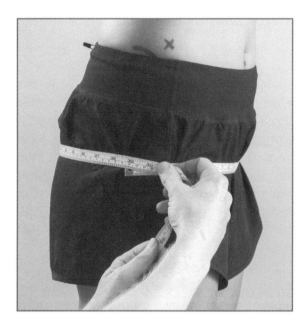

FIGURE 4.21   Hip (buttocks) circumference measurement.

Thigh Circumference

Ask the participant to stand with his right leg just in front of his left leg, with his weight on his left leg. Demonstrate this stance for the participant. Squat or kneel down to the right of the participant. Holding the zero end of the tape in your right hand, place the measuring tape around the midthigh at the point that is marked (+) in figure 4.13. Make sure the tape is positioned perpendicular to the long axis of the thigh and not to the floor. Once set, switch the zero end of the tape to your left hand and the rest of the tape to your right hand. Pull the tape lightly with your left hand until the appropriate tension is achieved (see figure 4.22). Hold the tape in place with your right hand. Record the measurement to the nearest 0.1 cm. (0.04 in.) (5).

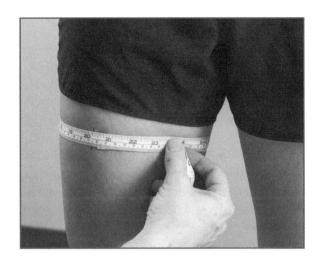

FIGURE 4.22 Thigh circumference measurement.

Calf Circumference

Ask the subject to sit on the end of a table with her lower leg hanging freely at a 90° angle to the thigh or to stand with weight on the left leg and the right foot on a platform so that the knee and hip are flexed to about 90°. Holding the zero end of the tape in your right hand, place the measuring tape around the calf at the point that is marked (+) in figure 4.15. Make sure the tape is positioned perpendicular to the long axis of the lower leg and not to the floor. Once set, switch the zero end of the tape to your left hand and the rest of the tape to your right hand. Pull the tape lightly with your left hand until the appropriate tension is achieved. Hold the tape in place with your right hand. Record the measurement to the nearest 0.1 cm (0.04 in.) (5,28) (see figure 4.23).

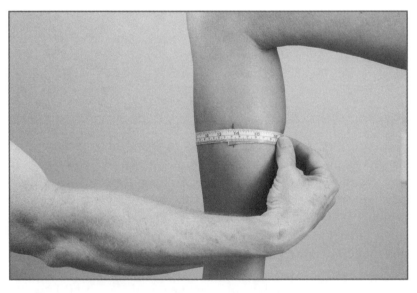

FIGURE 4.23 Calf circumference measurement.

Body Fat Measured From Circumference

Circumferences have been used as independent measures of adiposity or in combination with skinfolds to assess body composition. WC is often used to estimate abdominal adiposity. It is a good predictor of visceral adipose tissue and, in one study, accounts for 65% of the variance (34). However, in a commentary by Bouchard (35), it is clear that body mass index (BMI), fat mass (FM), and WCs are equally correlated with visceral fat and that WC is better correlated to total fat than to visceral fat.

Circumferences have also been found to effectively estimate body fat and FFM in children. Circumference equations have been shown to account for 55% to 83% of the variance in body fat and FFM estimates for children and youth. The circumferences used most often are abdomen, arm, waist, calf, and forearm (36,37). The association between circumferences and body fat in adults is similar to skinfolds and body fat, with SEEs of 3% to 4% fat (1).

Waist and hip circumferences and waist-to-hip ratios are significant predictors of cardiovascular disease (CVD) (38,39). In a study with 4,487 women aged 20 to 69 years without heart disease, diabetes, or stroke, WC and waist-to-hip ratio were found to be significant and independent predictors of CVD risk (39). Among nonobese women, increases in hip circumferences have been shown to significantly reduce the risk of CVD events (40,41). Considered together (but not as a ratio measure), waist and hip circumferences may also improve risk-prediction models for CVD and other disease outcomes (38), whereas WC alone is also often recommended.

Circumferences have been used by themselves to assess body fat using prediction equations with varying degrees of success. When compared with other field level methods, some studies report that circumference equations are more effective (36,42), and other studies suggest that circumference equations have less precision and accuracy (43,44). Although the SEEs of some of these prediction equations are adequate, they provide inconsistent results among investigators. Many of these circumference equations have not been cross-validated, were developed with homogeneous populations, and will have limited utility for other populations.

Circumference equations developed by Katch (45) designed to estimate body fat in college-age women were reported to have a measurement error of 2.5% to 4% fat. Cross-validations were not provided for the equations, and the equations appear to work best for individuals with a fat content of 20% to 30% because prediction errors increase for lean and obese individuals (45-48). When circumferences are included in prediction equations in combination with skinfolds to predict body fat, some investigators found they reduced the predication error (49), whereas other investigators found they provided no additional information (50).

Bioelectrical Impedance Analysis

Bioelectrical impedance analysis (BIA) is a body composition procedure that can be used to estimate body fat percent, body mass, and total body water (TBW) based on the electrical properties of tissue. The general principles were discovered as early as 1871, and by the 1970s, foundations of BIA were established, including those that underpinned the relationships between the impedance of a current and body composition. A variety of single-frequency BIA analyzers became commercially available in the 1980s.

The theory is that the volume of a conductor is related to its length (L) and its impedance (Z). Impedance of the body is affected by the specific resistivity and volume of the conductor. For the body, the conductor of the electrical current is the FFM, or, more specifically, the TBW. It can be shown that the volume of the conductor (fat-free body) can be predicted as follows:

$$V = p \, (L^2/Z) \tag{4.1}$$

where V = volume; p = specific resistivity of the fat-free body, or the water of the body; L = length of the conductor; and Z = impedance of the conductor.

It can be shown that impedance is the sum of resistance (R) plus reactance (Xc) of the body as follows:

$$Z = \sqrt{R^2 + Xc^2} \tag{4.2}$$

And because Xc is much smaller than R,

$$Z \sim R \tag{4.3}$$

Thus, the volume of a conductor can be estimated from its length and resistance.

$$V = p \, (L^2/R) \tag{4.4}$$

BIA is a practical approach to body composition assessment and similar to anthropometry in that it is safe, cost effective, mobile, convenient, and easy to use. The method involves passing an extremely low-strength electrical current through the body and measuring the resistance and reactance (impedance) to the flow of this current. BIA is based on conductivity differences between tissues (specific resistivity) because tissues with significant fluid levels (such as lean tissue) have less impedance than those that do not have large amounts of water, such as adipose tissue (51,52). The estimation of body composition using the BIA technique relies on the assumption that water as a proportion of FFM is known and is constant. This method also assumes the body is a cylinder of uniform circumference.

Although BIA is a recognized field method for estimating body composition, there is much variation among studies on its acceptable measurement accuracy. The accuracy of a BIA device depends primarily on the hydration status of the body. Also, the correct placement of the electrode is important. Dehydration is a recognized factor affecting BIA measurements because it causes an increase in the body's electrical resistance and can cause an underestimation of FFM, which results in an overestimation of body fat (51). Thus, the measurement accuracy of the technique is related to how well the research protocol ensures normal hydration in the subject.

The use of BIA to measure TBW and thus hydration level dates back several years, and many new advances are being made in that arena. Body fluid assessments are determined by BIA measures of resistance and impedance and are proportional to body water volume if body electrolyte status and hydration are normal. The science of BIA is based on two key concepts: the fact that the body contains water and conducting electrolytes and that impedance of a geometric system is related to conductor length, its cross-sectional area, and its signal frequency. When a current is passed through the body, the water-containing cells conduct the electrical current. Water is found both inside the cells, intracellular fluid (ICF), and outside the cells, extracellular fluid (ECF). At low frequencies, current passes through the ECF space and does not penetrate the cell membrane. At high frequencies, however, the current passes through both the ICF and ECF. Based on these concepts, a value for impedance can be calculated from a fixed-strength current being passed through the body, which is inversely proportional to the amount of fluid. By appropriate choice of signal frequency, this can be made specific for ECF or for total fluid determinations.

A detailed review of early research (1985-1990) with BIA and body composition is described by Lohman (1). For prediction of body water, SEEs ranging from 0.96 to 1.8 kg (2.1 to 4.9 lb) are found by Kushner and Schoeller (53), Davies et al. (54), and Schols et al. (55). Somewhat higher predictions are found by Van Loan and Mayclin (56) and Van Loan et al. (57) with SEEs of 3.2 kg (7.1 lb). In the review of this research, Lohman observed that the lower prediction errors were associated with a more accurate measure of body water using mass spectrometry rather than infrared spectroscopy (1).

The Valhalla study was designed to determine whether BIA estimates of percent fat and FFM are similar to skinfold estimates when a carefully designed protocol was followed among six laboratories. The results (1) indicated that SEEs of 2.9 kg (6.4 lb) FFM (3.7%) in men and 2.1 kg (4.6 lb) FFM (3.5%) in women were found for BIA, and similar results were found for skinfolds using the combined sample. Thus, the authors concluded that much of the variation in previous research was caused by methodological differences in protocol and measurement precision such that some authors found skinfolds to be a better approach (lower SEE) and other authors found BIA to produce a lower SEE.

PRACTICAL INSIGHTS

SEEs are always expressed in the unit of measure for the variable being measured. At the beginning of the chapter, acceptable SEEs were given for percent body fat, but these studies are focused on FFM in kilograms. Acceptable SEEs for FFM in kilograms are 3.5 kg for men and 2.8 kg for women; it is even better to aim for an SEE of about 2.5 kg. Any SEE larger than these numbers means that the method produces errors that are too large to be an acceptable body composition field method. Companies that want you to buy their products may want you to think their products are accurate, but the only way to know is to look at the validation studies where they report the SEEs comparing their products to a four-component model.

Bioimpedance spectroscopy (BIS) takes measurements at 256 different frequencies and uses mathematical modeling to calculate the resistance at zero and infinite frequencies (Rinf). These values are utilized to derive FFM and FM. The determination of impedance at zero frequency is highly significant because this value represents the impedance of the ECF alone, whereas establishing Rinf allows reliable prediction of the TBW (58).

BIS has been developed in the determination of under- or overhydration, which is critical in many disease states and to the medical profession in general. Phase angle can be calculated, which is the slope of the relationship between resistance and reactance, two characteristics of how the electrical current passes through the body (51). The multiple frequencies that are measured using BIS reflect intracellular versus extracellular volume, which relates to hydration status.

Compared with DXA and four-component models, both multifrequency BIA (MF-BIA) and single-frequency BIA (SF-BIA) accurately assessed changes in body composition with weight loss, but compared with SF-BIA, MF-BIA provided superior cross-sectional estimates of body composition. Foot-to-foot (such as scale-based BIA systems) and hand-to-foot BIA (scale-based systems with hand paddles) have been found to provide lower agreement with DXA and magnetic resonance imaging (MRI) for the assessment of whole-body composition in individuals than tetrapolar and eight-electrode BIA systems (59).

BIA requires an accurate measurement of height because the impedance to flow is affected by the length of the conductor. Therefore, the height–impedance index is used in many prediction equations for BIA. Prediction equations often include age and gender, and some models of BIA also use an estimate of physical activity level (chapter 7). In general, height/resistance^2 is the best single predictor of FFM and TBW. The addition of other anthropometric variables commonly used only slightly improved the prediction (60). Further, many equations have been published to predict body composition from BIA. Choosing the correct equation for the population being measured is necessary to achieve more accurate results (chapter 7). Impedance instruments include SF-BIA (see figure 4.24).

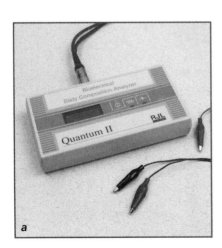 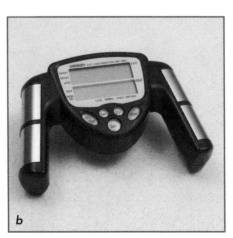 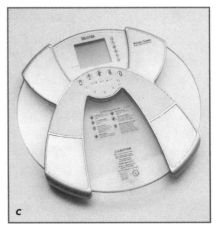

FIGURE 4.24   Single-frequency BIA systems: (*a*) hand-to-foot tetrapolar instrument, (*b*) arm-to-arm instrument, (*c*) leg-to-leg instrument.

SF-BIA is generally performed at a frequency of 50 kHz. At this frequency, the current passes through both the intracellular and extracellular fluid, and consequently, TBW may be calculated (see figure 4.25). However, as the current passes through both intracellular and extracellular compartments, variation in ICF and ECF is unable to be determined. SF-BIA relies on prediction equations and algorithms to calculate TBW and FFM. A single algorithm is not suitable for all subjects because all populations were not considered when the equations were developed. Arm-to-arm SF-BIA has been found to underestimate body fat in both men and women, but the procedures can effectively estimate body fat with adjustments (61). For this chapter, we focus on the different populations (e.g., children and adults). In chapter 7, different populations, including children and athletes, as well as racial and ethnic differences within the general population are explored (58).

FIGURE 4.25 Arm-to-arm SF-BIA procedure.

A 50 kHz leg-to-leg SF-BIA system was evaluated for accuracy of predicting body composition in the late 1990s by Nunez et al. (59). The system combined a digital scale that employs stainless steel pressure-contact foot pad electrodes for standing impedance and body weight measurements. Healthy adults were evaluated for electrode validity and potential for body component estimation. Pressure-contact foot-pad electrode–measured impedance was highly correlated with impedance measured using conventional gel electrodes applied to the plantar surface of both lower extremities; however, the mean impedance was systematically higher by about 15Ω for pressure-contact electrodes. The leg-to-leg pressure-contact electrode BIA system has overall performance characteristics for impedance measurement and body composition analysis similar to conventional arm-to-leg gel-electrode BIA and offers the advantage of increased speed and ease of measurement (see figure 4.26).

A tetrapolar (head-to-foot) impedance method (see figure 4.27) was evaluated for validity in 114 male and female participants, aged 18 to 50 years, with a wide range for FFM (34-96 kg [75-212 lb]) and percent body fat (4%-41%). Regression coefficients in the male and female regression equations were not significantly different. The results of the present study showed that the tetrapolar impedance method is valid and reliable and that it could be useful in field assessment of body composition among healthy people under steady-state conditions (51).

Excellent results predicting body composition were obtained using eight-electrode segmental MF-BIA by Bosy-Westphal et al. (62) (see figure 4.28). In this well-designed study using the four-component model as the reference method, the authors

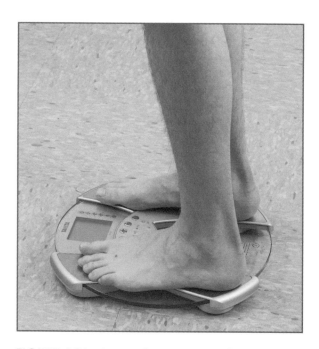

FIGURE 4.26 Leg-to-leg BIA procedure.

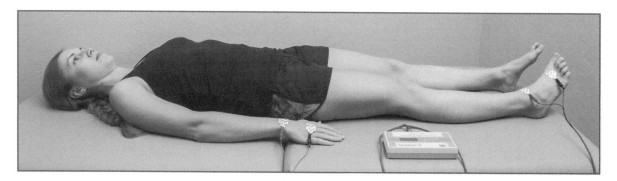

FIGURE 4.27 Hand-to-foot tetrapolar measurements.

found the whole-body resistive index (Ht^2/R_{50}) alone predicted FFM with a SEE of 3.2 kg (7 lb) and that, by adding to the multiple regression analysis the variables reactance, trunk resistance index, weight, gender, and age, the SEE decreased to 1.9 kg (4 lb). Similarly, the prediction of body water decreased from 2.3 to 1.4 kg (5-3 lb) by adding several variables, including total trunk index and trunk reactance index. This approach of accounting for trunk resistance improves the usefulness of BIA and implies the eight-electrode segment multifrequency approach may be classified as a laboratory method with lower SEEs than the SF-BIA approach (63-65).

Courtesy of Seca.

FIGURE 4.28   Eight-electrode MF-BIA system.

Currently, there are several BIA instruments that employ different procedures, varying from hand-to-foot to eight-electrode instruments with SF-BIA and MF-BIA systems. Also, there is BIS (see figure 4.29), with the analysis of the entire spectrum of resistances over a large range of current frequencies (66) by impedance (67).

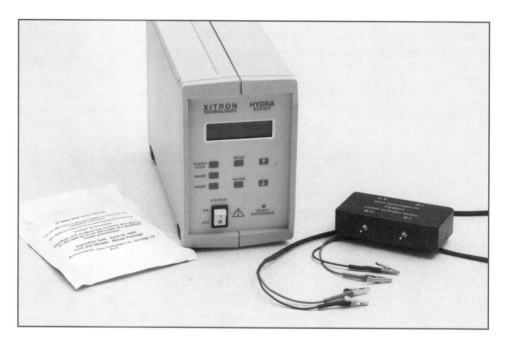

FIGURE 4.29   BIS measurement system.

Kyle and colleagues (68,69) published an excellent guideline paper on BIA assessments with different BIA instruments. They summarized a number of well-designed studies from 1985 through 2003 with healthy adults and provided a summary of prediction errors. Their studies show that segmental analyses and MF-BIA yield more accurate body composition equations than does SF-BIA whole-body analysis.

Both BIA and skinfolds have been used in national surveys and larger clinical trials to access body composition and changes in body composition. In 1994, a National Institutes of Health–sponsored consensus conference was held (Bioelectric Impedance Analysis in Body Composition Measurement), and it was concluded that reliable estimates of TBW can be provided by BIA under most conditions (70). One clinical trial used both skinfold and BIA methods to assess body composition changes and found that 3-year changes in percent fat in 663 children measured from 20 control schools along with 705 children from 21 intervention schools were 5.8% fat ± 5.0% fatter using BIA versus 6.3% fat ± 3.9% fatter using skinfolds for boys. For girls, similar results were found with a 6.0% ± 4.5% change using BIA and a 6.1% ± 3.3% change using skinfolds (71). The results are illustrated in scatter plots for both BIA and skinfolds (figure 4.30a and 4.30b). These plots indicate that there was considerably less variability in percent fat change from skinfolds.

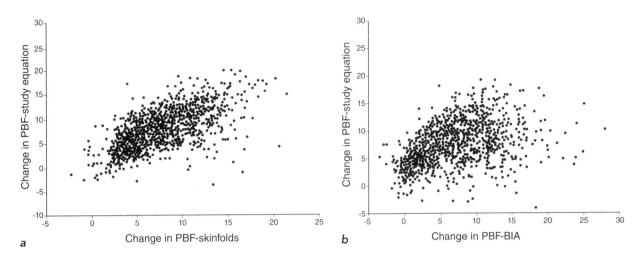

FIGURE 4.30 Scatter plot of boys' and girls' percent body fat estimates from (a) BIA and (b) skinfolds.

Reprinted by permission from T. Lohman et al., "Indices of Changes in Adiposity in American Indian Children," *Preventative Medicine* 37 Suppl 1(2003): S91-96.

A review of research from 1990 through 2010 by Elia (72) concluded that research with BIA, including all studies over the past 20 years, shows no consistent advantage by body composition estimates over height and weight alone. This conclusion differs greatly from the one found when reviewing studies where careful training and delineated protocol were used by experienced investigators.

Optimizing standardization of BIA body composition assessment procedures is essential. The following conditions should be employed when completing BIA body composition assessments to obtain the best results.

- The subject should not have exercised or taken a sauna within 8 h of the study.
- The subject should refrain from alcohol intake for 12 h prior to the study.
- The subject's height and weight should be accurately measured and recorded.
- The subject should lie or stand quietly during the entire test.
- For recumbent BIA tests, the subject should remove shoes and socks and lie supine for at least 10 min with the arms 30° from the body and legs not touching.
- The electrode sites should be cleaned with alcohol if the skin is dry or covered with lotion.
- The subject should not eat or drink within 4 h of the test.
- The subject should urinate within 30 min of the test.
- The subject should not take diuretic medications within 7 days of the test.
- Female subjects who perceive they are retaining water during their menstrual cycle should not be tested (7).
- Measurement should be performed in neutral ambient temperatures (i.e., ~72-85°F) after the subject has rested at that temperature for 15 to 20 min so that the skin temperature has stabilized.

In conclusion, the following recommendations are made relative to using BIA for FFM and body fat estimates.

- In healthy adults, SF-BIA equations must be validated in the population studied (e.g., race, age, sex specific) using a reference body composition method.
- Segmental MF-BIA may provide more accurate and generalized BIA equations for different populations.
- Technician training should include both instruction on how to use a BIA unit and how to properly monitor those factors that may affect a subject's hydration level.
- Any BIA or BIS manufacturer should be able to provide the SEE for their instrument for the estimate of TBW, FFM, or percent fat, which is especially important if the resistance and reactance values are not provided in the output. Units with high SEEs, and those using proprietary equations that do not allow the consumer to choose their own equations, should be avoided.
- With longitudinal changes, FFM and body fat changes can be assessed by BIA in subjects with normal hydration, provided a consistent protocol is followed. Also, repeated measurements must be made with the same BIA instrument because resistance can vary between BIA units.

Use of Weight and Height Indexes to Estimate Body Composition

A number of body indexes are currently circulating and being used by various constituencies. These indexes typically comprise height, weight, WC, and hip circumference calculations. These indexes were generated from body fat evaluated with a two-component (body fat and FFM) model and do not consider the individual body constituents that compose the FFM of the body. They are designed to be substituted for more precise estimates of body fat instruments because they are more accessible and less expensive. Their apparent measurement simplicity and conceptual relationships with established body fat measurements lead to questions about their validity (73,74). Body indexes most observed to estimate body adiposity include BMI; body adiposity index (BAI) (75); a body shape index (ABSI) (76); and a geometric-based index, the body roundness index (BRI) (77). Summaries of selected indexes are provided in the following sections. In general, various indexes have an SEE for predicting percent fat of 5% or greater.

Body Mass Index

BMI, or Quetelet index, is a measure of relative weight based on an individual's mass and height. BMI was devised between 1830 and 1850 by Apolphe Quetelet. It is defined as the individual's body mass divided by the square of her height, with the value universally being given in kilograms per unit of height in centimeters (equation 4.5). BMI is one of the most popular field methods for estimates of overweight and obesity. Because calculation requires only height and weight, it is inexpensive and easy to use for clinicians and the general public. BMI allows people to compare their own weight status to that of the general population (78). General guidelines for assessing overweight (BMI between 25 and 30) and obesity (BMI above 30) have been established in the adult population (78). The use of BMI to assess obesity and health risks has been criticized in scientific and lay publications because of its failure to account for body shape and inability to distinguish FM from lean and bone mass. BMI is used more successfully as an adequate measure of the mean obesity status of a given population, as opposed to individual obesity status. Therefore, for the individual, BMI should be considered a rough guide to body obesity status (78-80) because the typical SEE for obesity classification is 5% and considerably greater than skinfolds or BIA.

$$\text{BMI} = \text{weight in kilograms/height in meters squared} \qquad \textbf{(4.5)}$$

Waist Circumference

WC has emerged as a leading complement to BMI for indicating obesity risk. A number of studies have found that WC predicted mortality risk better than BMI. A World Health Organization report summarized evidence for WC as an indicator of disease risk and suggests that WC could be used as an alternative to BMI (81). In fact, WC is highly correlated with BMI, to the extent that differentiating the two as epidemiological risk factors can be difficult.

Body Adiposity Index

BAI is based on hip circumference and height in centimeters (equation 4.6) and has been proposed to be a predictor of body fat without the need for further correction for sex, race, or age. However, BAI does not provide consistent measures of body fat. In a study with 1,151 adults whose body composition was estimated by DXA, BAI, BMI, and waist and hip circumferences, the conclusion was that BAI, as an indicator of adiposity, is likely to produce biased estimates of percent body fat, with errors varying by sex and level of body fat (82,83). Bergman et al. (75) acknowledged that the relationship between BAI and percent fat is not linear. There is a different curvilinear relationship for each sex (83). Further, BAI is not as effective as other estimates of adiposity in predicting chronic diseases, including CVD (84).

$$\text{BAI} = (\text{hip circumference/height}^{1.5}) - 18) \qquad \textbf{(4.6)}$$

A Body Shape Index

With overweight and obesity reported to be near epidemic proportions in society, many are trying to determine the accuracy of assessment and the effect obesity is having on society. Because BMI is the basic tool most often used to estimate obesity worldwide, doubt as to the validity of BMI guidelines as indicators of obesity and related chronic health challenges exists. BMI does not distinguish between muscle and fat accumulation, and there is evidence that higher FM is associated with greater risk of premature death and higher muscle mass reduces risk. In an effort to address these concerns, ABSI, based on WC, BMI, and height and including sex-specific equations, was developed. ABSI was expected to be user friendly and account for factors that BMI does not and therefore be a more effective estimator of overweight and obesity (76). However, a study that compared anthropometric indexes of estimating obesity found that ABSI demonstrated the weakest correlations and lowest area under the curve for various cardiometabolic risk indicators. The authors state that ABSI has a significant association with abdominal adipose tissue and appeared to be more associated with premature death than WC or BMI. ABSI is a weak predictor for CVD risks and metabolic syndrome. Compared to BMI, ABSI seemed to have a lower association and to be a weaker index of hypertension and CVD (76).

ABSI is gender specific and therefore is calculated using two formulas.

$$\text{Women: WC/BMI}^{3/5} \times \text{Height}^{1/5} \tag{4.7}$$

$$\text{Men: WC/BMI}^{2/3} \times \text{Height}^{1/2} \tag{4.8}$$

Body Roundness Index

Based on limitations mentioned in the previous section, indexes are being proposed to replace BMI and WC as anthropometric estimators of adiposity and related chronic diseases. One such index is the BRI (85). This index was designed to bring together two major concepts in obesity phenotypes and risk assessment that have not been adequately addressed by BMI: predicting both total percent body fat and percent visceral adipose tissue using anthropometric body measurements (86) (equation 4.9). To accomplish this task, they viewed the human body as an ellipse to capture body girth in relation to height (body roundness). The girth captures waist and hip circumferences, and length accounts for height. First introduced in 1609 by the German astronomer Johannes Kepler to quantify the circularity of planetary orbits, the degree of roundness of an ellipse is characterized by a nondimensional value referred to as eccentricity (77). A cross-sectional study conducted in the rural areas of northeast China from January 2012 to August 2013 was completed on 5,253 men and 6,092 women with 1,182 participants (10.4%) suffering from diabetes mellitus (DM). Results showed neither ABSI nor BRI was superior to BMI, WC, or WHR at predicting the presence of DM (87). ABSI showed the weakest predictive ability, whereas BRI showed potential for use as an alternative obesity measure in the assessment of DM.

$$\text{BRI} = 364.2 - 365.5 \times \{1 - [\{\text{WC/}2\pi\}/(0.5 \times \text{height})]\}^{20.5} \tag{4.9}$$

Summary

Of the roughly 7.5 billion people in the world, 774 million are estimated to be obese, and 109.3 million of these are reported to be Americans. The health risks associated with overweight and obesity necessitate a simple, inexpensive, and effective protocol for body composition assessment. Field procedures such as skinfolds, circumferences, bioelectric impedance, and various indexes of height and weight are reviewed for their accuracy and practicality of use. While simple to measure, BMI and WC have limitations for individual assessment of adiposity. Evidence also suggests BMI and other height–weight indexes are inadequate proxies of body fatness in athletic populations (88). Skinfolds, circumferences, and BIA all provide more accurate estimates of body fat than height and weight. Use of these techniques with careful training, standardized protocols, and cross-validated equations can measure total body fat percentage within ±3% to 4%, and they are effective body composition assessment methods for both individuals and populations.

5

Assessing Measurement Error

Vinson Lee, MS

Leslie Jerome Brandon, PhD, FACSM

Timothy G. Lohman, PhD

LEARNING OBJECTIVES

After completing this chapter, you will be able to do the following:

- Define the types of measurement error
- Understand what comprises measurement error
- Understand precision and how it relates to error
- Learn how intra- and interobserver variation contributes to error
- Learn how absolute and relative technical error of measurement are calculated
- Observe how measurement error varies by body composition method

Ensuring the greatest possible accuracy of measurements is critically important when completing body composition assessments. There are many procedures used to assess body composition, and each procedure has technical, systematic, and procedural errors, making precise protocols and meticulous techniques necessary for obtaining the best measurements with the lowest possible error. Two key publications, the *Anthropometric Standardization Reference Manual* (1) and *Human Body Composition* (2), offer standardized procedures for anthropometry and various body composition methods so that more consistency can be obtained in the measurement of body composition.

In general, current technology and individual expert measurement skills allow for assessments with good precision. Precision, as defined in chapter 1, is the degree to which the same measurement under the same conditions and on the same participant produces the same results (also known as reproducibility or repeatability). A high precision means there is low variability in successive measures. To obtain high precision, assessment errors must be identified and controlled (3).

Types of Measurement Error

There are several types of measurement errors. The next three sections cover systematic and random errors, intra- and interobserver variability, and technical errors.

Systematic and Random Error

Systematic errors are measurement biases in one direction, which lead to measured values that are consistently higher or lower from the actual value. All measurements are prone to systematic errors, often of different types. Sources of systematic error may be imperfect calibration of measurement instruments, changes in the environment that interfere with the measurement process, and imperfect protocols or methods of observation. An example of systematic measurement error is when a protocol calls for a specific measurement of abdominal circumference and a different protocol is followed, leading to consistently higher or lower results. Systematic errors can be reduced by ensuring that all equipment is properly calibrated and taking meticulous measurements using standardized protocols.

Random errors are errors caused by the lack of predictability (uncertainty) that is characteristic of the measurement process and variation in the variable being measured. These errors fluctuate around the true value and, unlike systematic errors, are unavoidable. Random errors can be reduced by performing repeated measurements. Both systematic and random errors are inherent to the errors discussed in the next two sections.

Intra- and Interobserver Variability

Intraobserver variability is the error or difference obtained by the same investigator when completing the same assessment with the same participant using the same equipment and employing the same techniques. These errors are assessed two ways for this discussion: repeated assessments within the same day (within day) and assessments between days (interday). When assessments are made on different days, other sources of variation, such as hydration status of the participant, may contribute to intraobserver measurement error, whereas fewer sources of error are present when repeated assessments are performed within the same day.

PRACTICAL INSIGHTS

There are two kinds of errors that are important to understand. Systematic errors occur when a bias is introduced in the measurement, causing it to be consistently higher or lower than the actual value. An example of how this could occur is when a technician incorrectly reads a scale on a skinfold caliper as each line equaling 2 cm when each line really equals 1 cm. Random errors occur equally in both directions and do not cause a bias in the measurement. They are not consistently in one direction or another. Better training and more practice can reduce both kinds of errors. It is also recommended that your measurements be certified against a trained professional's measurements so that you are sure you are making accurate assessments.

Interobserver variability or error is a measure used to assess the degree of agreement among investigators. It provides a score that indicates the consensus among investigators, often called objectivity of the measurement. Assessment of interobserver variability is necessary because different investigators typically experience measurement error regardless of their efforts to ensure a high degree of precision. This is especially true with inexperienced investigators completing body composition measurements. To better understand and effectively compare body composition assessments of people from different research laboratories across the world, standardized procedures that yield minimum errors are needed.

Following standardized protocols should yield acceptable intra- and interobserver variation for a given body composition measurement method.

Absolute and Relative Technical Error of Measurement

Technical error of measurement (TEM) is used to estimate a lack of precision in the measurement results. Technical measurement error includes systematic and random errors. The total or absolute TEM is reported in the standard units of the variable measured, and the relative TEM is reported as a percentage. Using the mean of repeated measurements (two or three replicates) reduces random error caused by procedural inaccuracy (observer error) but not the systematic error caused by incorrect calibration of the measuring instrument or other sources of systematic bias, such as variation in measurement technique. Absolute TEM is assessed from many observations and subjects with the following equation, which was introduced in chapter 1:

$$\text{Absolute TEM} = \sqrt{\frac{\sum\left(D^2\right)}{2n}} \qquad (5.1)$$

where the numerator is the sum of the difference (D) between the two measurements squared for each subject measured and the denominator is the number (n) of subjects measured.

Relative TEM is assessed with the following equation:

$$\text{Relative TEM} = \frac{\text{TEM}}{\bar{x}} \times 100 \qquad\qquad \textbf{(5.2)}$$

where TEM is absolute TEM calculated by equation 5.1 and $\bar{x}$ is the mean of all measurements. Relative TEM is also known as the coefficient of variation (CV).

To better understand precision and how it is applied to assessment of error, the following sections will focus on TEM within observers (intraobserver) and between observers (interobserver) for various body composition measurement methods. Acceptable values of absolute and relative TEM will vary depending on the body composition component being assessed (4).

Intra- and Interobserver TEM/CV of Various Body Composition Measurement Methods

Laboratory and field measures of body composition, if performed carefully using standardized, accepted protocols by experienced investigators and technicians, have uniformly high reproducibility and relatively low error. The following sections and tables present TEMs and CVs that represent expected precision by method (4-29). Intra- and interobserver TEMs/CVs are not reported for all methods due to a paucity of available published data. Tables include both intra- and interobserver data when such data are available.

Laboratory Methods

The next several sections describe intra- and interobserver reliability for the four-component model and four laboratory methods.

PRACTICAL INSIGHTS

Variations among investigators have not been emphasized when teaching field methods, but this is an important part of a standardized training. It is not enough to follow a standardized protocol, although this is the foundation of training. One must have enough skill and consistency of measurement (intraobserver TEM) to be able to then decrease the variation among investigators to avoid systematic errors (interobserver TEM). With training, interobserver differences can be decreased to achieve satisfactory levels for the different methods. Finally, investigators must achieve a level of accuracy, which means they need to be compared to a technician who is known to be accurate. This process can take place as part of a certification process, which involves first learning the measurement protocols and practicing them, followed by becoming certified against a trained professional.

Four-Component Model

The four-component model for estimating body composition involves estimation of percent fat from measures of body density, body water, and bone mineral. Because of the greater number of measures and their associated errors involved, there was initial concern with the model's reproducibility because of propagation of measurement errors. However, several studies in adults and children have shown that the propagation of measurement errors is not a problem and that the reproducibility of estimates of percent fat from this reference method is similar to that of individual laboratory measures (5-9). Intraobserver CVs of 0.6% to 1.6% have been reported (table 5.1).

Table 5.1 Intraobserver Errors for Four-Component Model

Reference	Population	Intraobserver
Wells et al. (7)	Children	%Fat CV = 1.6% FM = 0.54 kg (1.2 lb) (TEM)
Friedl et al. (5)	Adults	%Fat CV = 1.1%
Withers et al. (8)	Adults	%Fat CV = 0.6%
Heymsfield et al. (9)	Adults	%Fat CV = 1.6%

%Fat = percentage of fat; FM = fat mass; CV = coefficient of variation; TEM = technical error of measurement.

Underwater Weighing

As the oldest of modern methods for measuring body density, underwater weighing (UWW) requires the subject to be comfortable with total submersion while exhaling as much air as possible. This method involves measuring body weight in air and body weight under water using a force transducer system, with residual lung volume measured by nitrogen washout, oxygen dilution, or helium dilution (10,11). Residual volume can be measured prior to, during, or after the measurement of UWW following a full or partial exhalation. Residual lung volume predicted from body size does not result in sufficiently accurate measures of body density. The procedure has a total TEM equivalent to approximately 0.3% to 1.6% fat (table 5.2). Of all error sources, the measurement of the residual lung volume contributes the most error. The procedure can have more error when used with children or adults uncomfortable under water, although acceptable reproducibility can usually be achieved with sufficient practice (9).

Table 5.2 Intraobserver TEM for Underwater Weighing

Reference(s)	Population	Intraobserver
Wells et al. (7)	Children	1.0 kg (2.2 lb) FM 2.9% fat
Friedl et al. (5) and Withers et al. (8)	Adults	TEM = 0.3% fat
Heymsfield et al. (9)	Adults	TEM = 0.0035 kg/L (0.0292 lb/gal) for Db TEM = 1.6% fat

FM = fat mass; TEM = technical error of measurement; Db = density.

Air Displacement Plethysmography

Body density can also be measured with air displacement plethysmography (ADP) using the BOD POD (COSMED, Rome, Italy), which is quick, automated, and more comfortable than UWW and accommodates a wider range of subjects, including children, the elderly, and disabled. The BOD POD system involves measuring body weight with an electronic scale, the air displaced by the body in the plethysmograph, and the average amount of air in the lungs during normal breathing. (See chapter 3 for a complete description of ADP.) In a review of the technique, Fields et al. (12) reported within-subject CVs of 1.7% to 3.7% for repeated measures of percent fat within the same day and 2.0% to 2.3% interday in adults (table 5.3). Thus, the TEM for ADP in adults is somewhat greater than the TEM for the estimation of body fat using UWW. Repeated measurement of percent fat CVs for air displacement plethysmography in children has not been reported (12).

Table 5.3 Interobserver CVs for Air Displacement Plethysmography

Reference	Population	Intraobserver
McCrory et al. (13)	Adults	%Fat CV = 1.7%
Fields et al. (12)	Adults	Within-day %Fat CV = 1.7% to 3.7% Interday %Fat CV = 2.0% to 2.3%

%Fat = percentage of fat; CV = coefficient of variation.

Total Body Water

Using the dilution technique, total body water (TBW) is measured by having individuals ingest deuterium, titrated water, or oxygen-18, allowing it to distribute throughout the body for several hours and then measuring the concentration of the substance in saliva, serum, or urine using an infrared spectrometer, a mass spectrometer, or a scintillation counter. CVs of 1% to 2% have been reported (5,7,8,14) for TBW and estimates of percent fat, indicating a high degree of reliability (table 5.4).

Table 5.4 Intraobserver CV/TEM for Total Body Water

Reference	Population	Intraobserver
Wells et al. (7) (deuterium dilution)	Children	FM = 0.27 kg (0.60 lb) (TEM) %Fat CV = 0.80%
Friedl et al. (5) (deuterium dilution) Intraassay CV for TBW	Adults	Based on 3 trials on different days = 1.0 L (0.3 gal), or ~1% TBW for %Fat CV = 1%, or 0.5 L (0.1 gal) (TEM)
Withers et al. (8) (deuterium dilution with saliva samples)	Adults	TBW = 0.25 kg (0.55 lb) (TEM)
Schoeller et al. (14) (deuterium dilution)	Adults	%Fat CV = 1%

CV = coefficient of variation; TEM = technical error of measurement; FM = fat mass; %Fat = percentage of fat; TBW = total body water.

Dual-Energy X-Ray Absorptiometry

Determination of total body composition from dual-energy X-ray absorptiometry (DXA) involves scanning the body with X-ray beams at high- and low-energy levels to determine fat, fat-free soft tissue, and bone mineral. In general, the reproducibility of this approach has been found to be high, with a relatively low error for measuring percent fat, fat-free mass (FFM) or fat-free soft tissue, and bone mineral. However, interpretation of reproducibility values from different studies and generalization across DXA manufacturers can be difficult because of the differences between DXA machine manufacturers' models, which may use different X-ray scan modes, and analysis software, which is continually being upgraded. In reviewing the literature, Toombs et al. (15) reported CVs ranging from 0.6% to 4.0% for percent fat and 0.4% to 1.3% for FFM (table 5.5). CVs for bone mineral content are less than 1%. Machine manufacturer, model, and X-ray beam type appeared to have little influence on reproducibility in contrast to their larger effects on accuracy where there are well-documented systematic differences among manufacturers. A report by Gutin et al. (16) found that the reliability of DXA in children is also high.

Table 5.5 Intraobserver CV/TEM for DXA Measurements

Reference	Population	Intraobserver
Gutin et al. (16) (Hologic QDR 2000)	Children	%Fat CV < 1%
Fuller et al. (17) (Lunar DPX)	Adults	FM CV = 3% FFST CV = 0.8% BM CV = 0.9%
Hind et al. (18) (Lunar iDXA)	Adults	FM CV = 0.82% %Fat CV = 0.86% FFST CV = 0.51%
Withers et al. (8) (Lunar DPX-L)	Adults	FM = 30 g (1.1 oz) (TEM)

CV = coefficient of variation; TEM = technical error of measurement; %Fat = percentage of fat; FM = fat mass; FFST = fat-free soft tissue; BM = bone mineral.

Further analysis of measurement precision for DXA across observers and interday was performed by Vicente-Rodríguez et al. (19), who evaluated the measurement errors for two observers from two labs in different cities. The TEM for intraobserver, interobserver, and interday was observed for percent body fat. Intraobserver TEM was between 0.5% and 1.1%. Interobserver and interday TEM ranged from 1.3% to 1.8% fat.

Field Methods

The following four sections cover intra- and interobserver variability for bioelectrical impedance analysis (BIA) and height and weight.

Skinfolds

There are many estimates of TEM for various skinfold sites following a standardized protocol. One such summary is the work of Ulijaszek and Kerr (4) where estimates of

TEM (intraobserver) for triceps skinfold (21 studies) vary from 0.1 to 3.7 mm (0.004-0.15 in.) ($\bar{x} = 0.84$) and for subscapular skinfold vary from 0.1 to 7.4 mm (0.004-0.3 in.) ($\bar{x} = 1.26$). The interobserver TEM for triceps skinfold varies from 0.2 to 3.7 mm (0.008-0.15 in.) ($\bar{x} = 1.1$) (19 studies) and for subscapular skinfold vary from 0.1 to 3.3 mm (0.004-0.13 in.) ($\bar{x} = 1.21$). In general, skinfold CV for interobserver error is about 10% for a given site, with a higher variation for suprailiac and abdominal skinfolds, which are more challenging to perform. Highly trained anthropometrists can attain a CV of 5.0% to 7.5% for most skinfold sites (20). Table 5.6 shows the TEM for triceps and subscapular skinfold sites in various populations.

Table 5.6 Intra- and Interobserver TEM for Triceps and Subscapular Skinfold Sites

Reference	Population	Intraobserver	Interobserver
Stomfai et al. (21)	Children (2-5 years)	Tri = 0.24 mm (0.009 in.) Sub = 0.17 mm (0.0067 in.)	Tri = 0.59 mm (0.023 in.) Sub = 0.32 mm (0.013 in.)
Stomfai et al. (21)	Children (6-9 years)	Tri = 0.24 mm (0.009 in.) Sub = 0.18 mm (0.0070 in.)	Tri = 0.61 mm (0.024 in.) Sub = 0.45 mm (0.018 in.)
Nagy et al. (22)	Adolescents	Tri = 0.23-0.75 mm (0.009-0.020 in.) Sub = 0.19-0.79 mm (0.007-0.031 in.)	Tri = 0.23-0.75 mm (0.009-0.020 in.) Sub = 0.19 to 0.79 mm (0.007-0.031 in.)
Harrison et al. (23)	Adults	Tri = 0.4 to 0.8 mm (0.016-0.031 in.) Sub = 0.9 to 1.2 mm (0.035-0.047 in.)	Tri = 0.8 to 1.9 mm (0.016-0.075 in.) Sub = 0.9 to 1.5 mm (0.035-0.059 in.)

Tri = triceps skinfold; Sub = subscapular skinfold.

Circumferences

The intra- and interobserver errors for circumferences are usually 0.5 to 1.5 cm (0.197-0.591 in.), depending upon subject size and protocol (table 5.7). The most challenging circumferences are in the trunk area, including hip, abdomen, and waist. Wang et al. (24) carefully studied four protocols for assessment of waist circumference and showed important absolute differences, with all sites being similar in reliability of measurement with CVs of 0.6% to 1.0% (24). (Refer to chapter 4 for a discussion on protocols for circumference measurements.)

Bioelectrical Impedance Analysis

The intra- and interobserver errors for bioelectrical impedance analysis (BIA) are relatively low with a 1Ω or 2Ω variation within and between observers using a standardized protocol (table 5.8). When measuring resistance in the prone position, it is essential to wait 3 to 5 min to obtain more stable readings. The largest source of variation between investigators is related to variation in the placing of electrodes on the distal locations of hand and foot. Following the standardized protocol for placing

Table 5.7   Intra- and Interobserver TEM for Circumference Measurements

Reference	Population	Intraobserver	Interobserver
Stomfai et al. (21)	Children (2-5 years)	Waist = 0.79 cm (0.311 in.) Hip = 1.2 cm (0.472 in.) Arm = 0.22 cm (0.087 in.)	Waist = 0.5 cm (0.197 in.) Hip = 0.5 cm (0.197 in.) Arm = 0.25 cm (0.098 in.)
Stomfai et al. (21)	Children (6-9 years)	Waist = 0.41 cm (0.161 in.) Hip = 0.31 cm (0.122 in.) Arm = 0.21 cm (0.083 in.)	Waist = 0.62 cm (0.244 in.) Hip = 0.63 cm (0.248 in.) Arm = 0.3 cm (0.118 in.)
Callaway (25)	Adolescents	Waist = 1.3 cm (0.512 in.) Hip = 1.2 cm (0.472 in.)	Waist = 1.6 cm (0.630 in.) Hip = 1.4 cm (0.551 in.)
Verweij et al. (26)	Adults	Waist = 1 cm-9 cm[a] (0.394-3.543 in.)	Waist = 1 cm-15 cm[a] (0.394-5.906 in.)
Callaway (25)	Adults	Arm = 0.1 cm-0.4 cm (0.039-0.157 in.)	Arm = 0.3 cm (0.118 in.)

[a]Ranges for intra- and interobserver TEM reported by Verweij et al. (26) in a systematic review of studies of varying methodological quality. Intra- and interobserver TEM was 1 cm to 6 cm (0.394-2.362 in.) in the better-designed studies.

Table 5.8   Intraobserver CV/TEM for Bioelectrical Impedance Analysis

Reference	Population	Intraobserver
Schaefer et al. (27)	Children	%Fat CV = 0.4% FFM CV = 1.23%
Nagy et al. (22) (resistance, ohms)	Adolescents	Ohms = 1.3 (TEM)
Lohman (28)	Adults	%Fat CV = 2.4% %Fat CV = 2.1%

CV = coefficient of variation; TEM = technical error of measurement; %Fat = percentage of fat; FFM = fat-free mass.

the electrodes can minimize this source. Variation in hydration state, skin temperature, time of meal intake, and time of exercise can all add to variation in resistance if a standardized protocol is not followed. Precise measurements do not always ensure accuracy (see chapter 1).

Height and Weight

Both height and weight can be reliably assessed following standardized protocols (table 5.9). There is diurnal variation in weight and height, so it is important to standardize the time of measurement for many research studies. The wide range in height and weight interobserver TEM found by Ulijaszek and Kerr (4) indicates that, although measurers were trained using a standardized protocol, precision varied across studies. For certification purposes, the relative CV/TEM should be less than 1% for both height and weight.

Table 5.9   Intra- and Interobserver TEM for Height and Weight (Males and Females)

Reference	Population	Intraobserver	Interobserver
Height			
Stomfai et al. (21) Crespi et al. (29)	Children (2-5 years)	0.0017 m (0.0669 in.) —	0.0026 m (0.1024 in.) 0.001 m (0.039 in.)
Stomfai et al. (21)	Children (6-9 years)	0.0022 m (0.0866 in.)	0.0026 m (0.2014 in.)
Ulijaszek & Kerr (4)	Adults	0.0038 m (0.1496 in.) ($\bar{x}$ of 19 studies) (Range = 0.001-0.013 m [0.039-0.512 in.])	0.0038 m (0.1496 in.) ($\bar{x}$ of 21 studies) (Range = 0.002-0.008 m [0.079-0.315 in.])
Weight			
Stomfai et al. (21) Crespi et al. (29)	Children (2-5 years)	0.05 kg (0.11 lb) —	0.13 lb (0.06 kg) 0.2 lb (0.1 kg)
Stomfai et al. (21)	Children (6-9 years)	0.06 kg (0.13 lb)	0.15 lb (0.07 kg)
Ulijaszek & Kerr (4)	Adults	0.17 kg (0.37 lb) ($\bar{x}$ of 6 studies) (Range = 0.1-0.3 kg [0.22-0.66 lb])	1.28 kg (2.82 lb) ($\bar{x}$ of 12 studies) (Range = 0.1-4.1 kg [0.22-9.04 lb])

Reducing Error Associated With Field Methods

There are a variety of field methods available for the measurement of body composition outside of the laboratory or clinical setting; however, the training protocols and certification criteria for these field methods are not well standardized. Measurements made in the absence of standardized training protocols and certification criteria may result in incorrect measurement technique, leading to increased variability and error (30).

Improved measurement evaluation in the field of body composition assessment is an essential part of its successful application with better training and certification. The technical error is composed of both a systematic and a random component. It is important for the purposes of certification to test for both the systematic and total technical errors. The systematic component can be calculated from the difference between means in each skinfold site comparing expert and novice. In general, experts have years of experience with measuring and researching a given body composition method.

Criteria for acceptability require the means to be within 10% for each site. The total relative technical error for most skinfold sites is also 10%. Exceptions are for suprailiac and abdominal skinfolds where a relative technical error of not more than 12% to 15% is recommended.

The International Society for the Advancement of Kinanthropometry (ISAK) has created an accreditation scheme (31) that ensures that all certified measurers use exactly the same measurement technique to help eliminate interrater variability and decrease error. The foundation of the accreditation is the objective maintenance of quality assurance, which requires that every measurer meet TEM criteria in the practical examination to be completed at the end of the ISAK accreditation course. The practical examination is followed by the measurement of 20 subjects to demonstrate that the measurer can meet further TEM criteria and show satisfactory repeatability of the measurements. Intertester relative TEM targets of 10.0% for each skinfold and 2.0% for all other measures must be met (31).

ISAK provides a four-level accreditation program, with each level more rigorous than the previous. Each level is directed at a particular measurement goal, and measurers are not required to meet the criteria of all four levels. Level 1 is designed for the evaluators who have little ongoing requirements for more than the measurement of height, weight, and skinfolds. Level 2 is designed for the technicians who wish to offer their subjects a more comprehensive range of measurements and requires demonstration of adequate precision in 3 base measures, 8 skinfolds, 9 segment lengths, 13 girths, and 7 bone breadths along with a broad understanding of the theory of anthropometry and its interpretation. Level 3 (instructor level) and level 4 (criterion level) are designed for only those anthropometrists who wish to engage in the training and accreditation of levels 1 and 2 anthropometrists.

Summary

Assessment errors must be identified and controlled to ensure that body composition measurements have high precision. Types of assessment errors include systematic errors (repeatable/reproducible and avoidable), random errors (not repeatable and unavoidable), and intra- and interobserver and technical errors. Intraobserver error is the difference in measurement values found when the same investigator or measurer performs repeat measurements on the same participant using the same equipment. Interobserver error is used to assess agreement between different investigators or measurers. TEM is calculated as both absolute and relative technical errors. Absolute TEM is reported in the units in which the measurement is reported, whereas relative TEM is reported as a percentage and is also known as the coefficient of variation (CV).

Systematic and random errors can be reduced by making sure equipment is properly calibrated, using standardized protocols, and performing repeated measurements. Intra- and interobserver errors and TEM are subject to systematic and random errors; therefore, reducing these errors will also reduce intra- and interobserver errors and TEM.

6

Estimation of Minimum Weight

Timothy G. Lohman, PhD
Kirk Cureton, PhD, FACSM

LEARNING OBJECTIVES

After completing this chapter, you will be able to do the following:

- Define minimum weight and discuss why athletes should not compete at less than minimum weight

- Identify the strengths and weaknesses of laboratory and field methods that can be used to estimate body composition to determine minimum weight

- Summarize the recommended practical approaches for determining minimum weight in athletes and their accuracy

The definition of minimum weight (MW) is derived from evidence that there is an increased risk of health problems in those individuals who become so lean that the amount of body fat is insufficient to foster optimal growth in children and adolescents and for maintenance of the musculoskeletal system in adults. MW is defined as the lowest weight that can be maintained indefinitely without adverse effects on health and performance. It has operationally been defined as 5% to 7% body fat in older teenage boys and men and 12% to 14% body fat in older adolescent girls and women (1,2).

In 1997, three college wrestlers died while engaged in rapid weight-loss programs prior to competition. Following the deaths, the National Collegiate Athletic Association (NCAA) implemented a mandatory minimum wrestling weight program, preventing wrestlers with an estimated body fat of less than 5% from competing (3), and soon high schools throughout the United States followed suit. In the United States, the National Federation of High Schools established a rule that required each state's high school association to develop a specific weight-control program that would discourage excessive weight reduction, including hydration testing, preceding a body fat assessment and an MW class based on 7% fat or greater for males and 12% fat or greater for females (4).

To estimate MW, an accurate estimate of body fat is needed. If we assign 5% fat for men and 12% for women, then MW for males equals

$$\frac{\text{FFM (kg)}}{0.95} \qquad \textbf{(6.1)}$$

and MW for females equals

$$\frac{\text{FFM (kg)}}{0.88} \qquad \textbf{(6.2)}$$

where fat-free mass (FFM) equals (5)

$$\text{Body weight} - \frac{\text{Body weight (\% Fat)}}{100} \qquad \textbf{(6.3)}$$

If percent body fat can be estimated with a standard error of estimate (SEE) of ±3%, then the corresponding error in FFM and MW is ±2.1 kg (±4.6 lb). For a SEE of ±2% fat, the error in FFM is decreased to ±1.4 kg (±3.1 lb).

Behnke (6) proposed that a fat content of 2% to 3% in the lean body mass is essential for life in males. For females, Behnke (6) added sex-specific essential fat in mammary and other tissues. MW in females was estimated to be about 5 to 7 kg (11-15 lb) above lean body mass (7). Katch et al. (8) proposed a model of fat distribution in females of 9% (5% sex-specific plus 4% essential fat). Below 12% for female athletes is now well accepted as the MW level (9).

A review of the health risks associated with body weight below the minimum by Sundgot-Borgen et al. (10) indicates common practices used to achieve low body weight include extreme dieting, frequent weight fluctuations, fasting, dehydration, purging, and excessive physical training for many athletes in weight-sensitive sports. Eating disorders, distorted body image, menstrual dysfunction, and loss of bone mineral often result from these practices.

PRACTICAL INSIGHTS

Estimating minimal weight has wide application to athletes, eating disordered populations, and those with chronic diseases. This field was pioneered in wrestlers where "making a weight class" resulted in high school and college wrestlers engaging in unsafe weight loss practices. Minimal weight is also helpful in long-term weight loss studies where healthy goals can be set for the individual to avoid excess loss of fat. Knowing the total adult fat-free mass of the individual provides a reference base throughout the life cycle to be maintained or increased depending on the client's goals. The aim is to achieve a weight that is healthy for the individual given their sport or life goals. Once someone falls below this minimal weight, health status deteriorates and death can occur.

The purpose of this chapter is to introduce the concept of MW for athletes and to discuss why it is important and how it can be measured. Laboratory and field methods for estimating body composition used to determine MW are reviewed, and practical recommendations for determining MW for athletes are presented.

Estimating Minimum Weight in Wrestlers

Estimating MW in wrestlers is especially significant in the history of applied body composition assessment to the athletic population. Because of the unhealthy weight-loss practices that are commonly used in sport, special efforts have been made to establish MW procedures for each wrestler at the beginning of the season to reduce the behavior of unhealthy weight-loss practices. Researchers and administrators at the University of Iowa and University of Wisconsin began exploring the development of MW protocols beginning in the 1960s.

The publication of the Midwest wrestling study (11), from the universities of Iowa, Illinois, Minnesota, Nebraska, and Ohio, resulted in a breakthrough study that was applied first throughout the state of Wisconsin (1991) and then expanded state by state to provide a practical innovative approach to MW estimates for wrestlers. The publication included obtaining support to estimate MW in all high school wrestlers from wrestlers, parents, administrators, health and medical personnel, and professionals of critical governing bodies.

Laboratory Methods for Estimating Minimum Weight

Laboratory methods for estimating body composition were discussed in chapter 3. In theory, any of the laboratory methods based on a chemical model that permits calculation of the FFM can be used to estimate MW, such as body densitometry or hydrometry, based on a two-component model; dual-energy X-ray absorptiometry (DXA), based on a three-component model; or three- or four-component models that combine the measurement of body density with measurements of bone mineral from DXA and/or body water. More elaborate five- and six-component models that involve neutron activation analysis also can be used for estimating MW. Other laboratory methods, computed tomography scanning, or magnetic resonance imaging (extensive analysis for total body fat) are not easily used for estimates of total fat and thus for estimating MW. In practice, estimates of MW in wrestlers using laboratory methods have been made from densitometry, DXA, and a four-component model (11-13).

Laboratory measures may provide more accurate estimates of MW than field measures. The total error in estimating MW using laboratory measures is from technical measurement error and biological error associated with individual variability about assumed values underlying the method. Estimates of MW from field measures have additional error associated with a regression equation used to predict a criterion body composition measure from simpler field tests. The accuracy of laboratory measures depends on carefully standardizing the conditions under which the measurements are performed and on using equipment and procedures known to be reliable and accurate by comparison with a reference method (e.g., four-component model). All of the laboratory measures, except the four-component model, make assumptions about the water content of the body, so standardizing the hydration state, prior diet, and acute exercise is essential. Although field measures will normally be used in most practical situations, when available and high accuracy is paramount, laboratory measures may be the preferred approach to obtaining estimates of MW.

Densitometry

Densitometry involves the measurement of the body density (body weight/body volume). Body volume has historically been measured using Archimedes' principle and hydrostatic weighting, but in recent years, body volume also has been measured using air displacement plethysmography (ADP) (see chapter 3). Hydrostatic weighing involves considerable cooperation from participants, requiring them to submerge completely under water, maximally exhale, and remain still. The procedure is difficult for children and adults uncomfortable in completely submerging. ADP is simpler and more suitable for a wider range of participants. Both hydrostatic weighing and ADP require measurement of the residual lung volume, which is technically difficult and involves expensive equipment. For many years, body densitometry using hydrostatic weighing was considered a reference method, providing the most accurate estimates of body composition available from an indirect method. Therefore, most of the practical field methods that have been developed for estimating MW from anthropometric measures in wrestlers have been validated against estimates of body composition from body density determined using underwater weighing (11,14). Some of these methods now have been cross-validated against estimates from DXA and a four-component model (12,15).

Estimates of body composition from body density based on a two-component model have limitations. As discussed in chapters 1 and 3, all estimates of body composition based on a two-component model are limited by the assumption that the makeup of the FFM or the concentration of a substance within the FFM is constant (5). In the case of body densitometry, in adults over the age of 20, it is assumed that the density of the FFM is 1.1 g/cm^3 (0.04 lb/in.^{3}) and that the FFM comprises 73.8% water, 19.4% protein, and 6.8% mineral. In children and adolescents, the proportion of water comprising the FFM is higher, and the proportion of mineral is lower, resulting in a density of the FFM (D_{FFM}) lower than 1.1 g/cm^3 (0.04 lb/in.^{3}). The density of the FFM increases progressively with age until it reaches the adult value at about age 20 years (16). In theory, it is important to take into account the lower D_{FFM} in estimating MW in adolescent athletes by using adjusted equations for estimating body fatness and calculating FFM from body density. Thorland et al. (11), in their comprehensive study, have developed equations for estimating MW from anthropometric measures in high school wrestlers using body composition estimates from hydrostatic weighing as the criterion and adjusting for the assumed lower D_{FFM}.

For individuals of any age, considerable individual variability exists in the portions of water, mineral, and protein comprising the FFM, resulting in an error in estimating FFM and MW from body density (17-19). Lohman (5) estimated that one-half of the error in estimating MW for high school wrestlers using his skinfold equation for predicting body density was likely related to individual variability in the composition and density of the FFM, which are assumed to be constant when body composition is estimated from body density. Variability in the water content of the FFM is the greatest source of biological error in estimating body composition from body density (17). Even when the state of hydration is normal, the water content of the FFM may not be what is assumed in very muscular athletes. Modlesky et al. (20) found that the greater water content of the FFM associated with high muscularity reduced the density of the FFM and caused overestimates of percent fat from body density in bodybuilders.

To determine whether estimates of MW from hydrostatic weighing were valid in collegiate wrestlers, Clark et al. (15) cross-validated estimates of MW from body density determined by hydrostatic weighing and the Lohman skinfold equation predicting body density (17) against estimates from a four-component model in which body density, body water, and body mineral were measured. They found estimates of MW from hydrostatic weighing and skinfolds agreed closely with the four-component model reference method (total errors of ±1.3 and ±1.7 kg [±2.9 and ±3.7 lb]). The strong agreement between the estimates from hydrostatic weighing based on a two-component model and those from the four-component model implied that the assumed density and composition of the FFM used in the equation to convert body density to fatness were appropriate for their group of collegiate wrestlers. They concluded that estimates of MW from body density using the hydrostatic weighing or the Lohman skinfold equation (11,21) were appropriate for collegiate wrestlers.

Dual-Energy X-Ray Absorptiometry

DXA has become widely used in hospitals, clinics, and universities as a laboratory method for estimating body composition because of its speed, convenience, and high precision (see chapter 3). The method is based on a three-component model and measures bone mineral, fat, and fat-free soft tissue. The sum of the bone mineral and fat-free soft tissue is the FFM, which can be used to estimate MW. An advantage of DXA for estimating MW is that, unlike other laboratory methods, DXA is minimally influenced by changes in body water (22,23). The widespread use of DXA for recommending MW to athletes is limited by its expense, the need for a trained and certified technician, and issues related to its validity (24). Although highly reproducible (~0.5%-1% error for repeated measurements of FFM and percent fat), estimates of body composition from DXA vary among instruments from different manufacturers, different models from the same manufacturer, and scan mode and software version (25). Measurements from very tall or wide individuals who do not fit on the scanning bed may also be problematic and require customized approaches (26,27).

A large number of studies have validated estimates of body composition from DXA against estimates from a four-component model, the current gold standard for body composition measurements. The results have been inconsistent, and some studies have reported large individual differences (25,28,29). Although most studies have found the error in DXA estimates of percent fat to be low, between 2% and 4%, there exists a tendency for underestimates of percent fat in leaner individuals and overestimates in fatter individuals (25). This trend is evident in some studies validating DXA percent fat and FFM estimations in athletes (30,31) but not in other studies of athletes (12,13,29,30). Also, small, individual changes in body composition of athletes may not be accurately tracked by DXA (29). A comprehensive study found that a widely used densitometer, the Hologic QDR 4500, overestimates FFM and underestimates fat mass by 5%, compared to estimates from a four-component model (28). The report recommended correcting all values for FFM and fat mass from this Hologic densitometer by 5%. Although the literature on the validity of DXA indicates that some caution is needed in employing body composition estimates from DXA to avoid underestimates of MW that could jeopardize health, especially in very lean individuals, studies reviewed below indicate that some Hologic, Lunar, and Norland DXA instruments do provide highly accurate MW estimates in collegiate and high school wrestlers.

Clark et al. (12) compared MW estimates from FFM obtained by DXA using a Norland XR-36 densitometer with estimates from hydrostatic weighing in 94 high school wrestlers. They found close agreement between the two estimates of MW with no significant mean difference (0.6 kg [1.3 lb]), a high correlation (0.98), a very low total error (1.9 kg [4.1 lb]), and no systematic bias across weight classes. The total error was lower than that reported in other studies for estimates of MW from field measures, including skinfolds, bioelectrical impedance analysis (BIA), and infrared reactance. They concluded that DXA provided an accurate estimate of MW in high school wrestlers under the conditions of their study.

Hydrometry

Body composition can be measured by measuring the total body water using the dilution principle, assuming that a constant percentage of the FFM is water (see chapter 3). A known amount of an isotope of water (deuterium, tritium, or oxygen-18) is consumed; allowed to disperse throughout the body for several hours; and followed by a measurement of the isotope concentration in blood, saliva, or urine (33). The advantages of the method are that it is easy to administer, requires little action from subjects, and is highly reliable (1%-2% technical error for mass spectroscopy) and accurate (total error 2%-4%) if the appropriate FFM hydration constant is used. A constant of ~73% is used in adults of all ages, with higher percentages used in individuals under age 20 years (16). The disadvantages of the method are the expense of stable isotopes and analytical equipment; lack of availability of stable isotopes; expert technical skills required for isotope analysis; time required for isotope equilibrium; and the sensitivity to acute or chronic changes in body water from exercise, illness, or chronic disease. Measurement in a state of normal hydration is essential. Because of the difficulties of the method, it has not been widely used as a standalone method to estimate body composition and MW. However, total body water has been used to estimate body composition in athletes and nonathletes in conjunction with estimates of body composition in multicomponent models because it is one of the measures required in densitometric three- and four-component models (20,30,31). In those studies, estimates of FFM and percent fat from body water agreed closely with those from a four-component model. A simple way to measure hydration in the field is a urine test of specific gravity not to exceed 1.025 g/cm^3 (0.037 lb/in.^{3}). This test (urine refractometer or urinometer/hydrometer) is required by the National Federation of High Schools to precede the body fat assessment (4).

Multicomponent Models

Estimation of body composition using a four-component model that involves measurements of body density, body water, and bone mineral or five- and six-component models that involve neutron activation analysis provide the most accurate estimates of body composition, with a precision and accuracy of estimating body fat of 1% to 2.5% error (see chapter 3). These methods are currently considered reference methods, which means they are used to validate other laboratory and field methods (24). The four-component model is the most widely used and addresses limitations of simpler laboratory methods, such as body densitometry based on a two-component model, by providing direct measures of body water and bone mineral. Direct measurement of body water is important because water is the most variable component of the FFM and contributes the most to the biological error estimating body fat from body density (17). Measurement of bone mineral can be important in estimating body composition of adolescents, who have incompletely ossified skeletons (16). Although use of multicomponent models is not a practical approach to estimating MW because of the expense, time, technical expertise, and lack of accessibility, it is important in checking the validity of other laboratory and field methods. At this point, only two studies (13,34) have used the four-component model to assess the validity of MW estimates in athletes from other laboratory and field methods. The four-component model study confirmed that hydrostatic weighing and Lohman's skinfold equation (35) are valid approaches for measuring MW in wrestlers. Evans et al. (34) used the four-component model to develop a skinfold prediction equation to estimate percent fat in college athletes. The equations developed can be used with Black and White male and female athletes to estimate MW (chapter 7).

Ultrasound

For estimating MW in the athletic population, the ultrasound method offers a more precise and accurate assessment of subcutaneous fat thickness than does the skinfold method (see chapter 3). The advantages include avoiding fat compression, individual skin thickness variation, and measurement variation associated with a double thickness of fat in the skinfold. Recent advances in ultrasound have led to software that determines the fat borders semiautomatically and measures multiple thickness values in each image (36,37). The possibility that MW can be estimated with a SEE of 1.5 kg (3.3 lb) or less depends on the relation of subcutaneous fat measured by ultrasound to total body fat in the athletic population. A large validation study is needed with an accurate reference method (i.e., four-compartment model) used to measure total body fat. A set of standardized ultrasound fat thickness sites, such as described by Störchle et al. (38), based on the work of Müller et al. (36,37), can be used to evaluate this method.

Field Methods for Estimating Minimum Weight

In nonlaboratory settings, such as schools and athletic departments in colleges and universities, simpler, less expensive field methods will be used to measure body composition to estimate MW in athletes. These methods are less accurate than laboratory methods but provide a practical approach to determine MW.

Skinfold Method

The work of Sinning (14) in college athletes validated skinfolds as a method of estimating MW. In 1984, Sinning and Wilson (39) developed skinfold equations for college female athletes, and, in 1985, Sinning et al. (40) developed them for males.

From the work of Sinning (14) in adult wrestlers, it is evident that the percent fat and FFM can be estimated more accurately using skinfolds than using skeletal widths. Prior to this research, skeletal widths were used to estimate FFM (41). In the classical work of Behnke and Wilmore (7), the derivation of the relation between skeletal widths and lean body mass is explained, and much of the early research is described. Tcheng and Tipton (42) also used this method to estimate MW in high school wrestlers. Using skinfolds improved the prediction of FFM and MW from SEEs of 4.0 kg (8.8 lb) for skeletal widths to SEEs of 2.0 to 3.0 kg (4.4-6.6 lb) for skinfolds.

The best equations for estimating MW using skinfolds in high school male wrestlers were published by Thorland et al. (11) in an interuniversity study of 806 participants. The details of that study are summarized in a monogram by Lohman (21). One of the equations recommended is

$$D = 1.0982 - 0.000815 \, (\Sigma \, 3 \text{ skinfolds}) + 0.00000084 \, (\Sigma \, 3 \text{ skinfolds})^2 \qquad \textbf{(6.4)}$$

This equation was developed by Lohman (43) by combining the work of others (14,44) to develop a generalized equation for young male adults using the triceps, subscapular, and abdomen skinfold sites. This equation is now widely used in high school wrestlers throughout the country. The sites and methods have been carefully described by Thorland et al. (11).

PRACTICAL INSIGHTS

Many skinfold equations are developed from one sample of participants by one investigator in one laboratory. As a result, they are not generalizable to other populations of people. A generalizable equation is one that is accurate in many groups of people rather than just one group. The unique aspect of the interuniversity study was the cooperation of five laboratories following the same protocol and standardization of methodology. The population-specific equation developed has been found to be highly generalizable to high school wrestlers across the country over the past 25 years. This is valuable because there are not multiple equations that were developed separately by multiple investigators published in the field; instead there is just one that can be used.

For female athletes, the skinfold equations of Jackson, Pollack, and Ward (45) using the sum of four skinfolds (triceps, abdomen, suprailiac, and thigh) have been shown to be valid (39).

BIA Method

The use of BIA as a field method to estimate MW in wrestlers has been investigated for over 25 years as an alternative field-based method (46). However, the comparability of BIA and skinfold approaches has been questioned (47).

Variation in the measurement protocol, type of BIA equipment used, and the criterion method selected makes it difficult to arrive at definitive recommendations. Evidence was presented in chapter 4 to show that, under standardized conditions in the young adult population, both skinfolds and whole-body BIA measurements can estimate percent fat with a SEE of 3.5% fat.

In a review article, Moon (48) indicates that investigators using BIA in athletes to assess body composition have yet to develop a generalized athletic-specific BIA equation based on a multicomponent reference method. Moon predicts that the BIA method could be a valid tool for athletes if a generalized equation were developed. Until generalized, athletic-specific BIA equations are developed using a multicomponent approach, he recommends using the equations in table 6.1 (45).

BIA Method in Athletic Populations

Clark et al. (13) validated estimates of MW from laboratory methods (hydrostatic weighing and DXA) and field methods (skinfolds and leg-to-leg BIA) against estimates from the four-component model in 54 collegiate wrestlers. They found close agreement between the five estimates of MW, with no significant mean differences (0.3 kg [0.7 lb] range of means), high correlations with the criterion (0.92-0.98), and no systematic bias across the range of MW. Total errors were excellent (<2.0 kg [<4.4 lb]) for hydrostatic weighing (1.3 kg [2.9 lb]) and skinfolds (1.7 kg [3.7 lb]) and only slightly higher and very good for DXA (2.2 kg [4.9 lb]) but too high to be useful for

Table 6.1   Recommended BIA Equations

Author(s)	Equations
Lukaski et al. (49)	FFM = 0.734 (Ht^2/R) + 0.116 + Wt + 0.096 + Xc + 0.878 × 1 (if subject is male) − 4.03
	FFM = 0.734 (Ht^2/R) + 0.116 + Wt + 0.096 + Xc + 0.878 × 0 (if subject is female) − 4.03
Lohman (5)	Male and female children 8-15 years: FFM = 0.620 (Ht^2/R) + 0.210 Wt + 0.100 Xc + 4.2
	Females 18-30 years: FFM = 0.476 (Ht^2/R) + 0.295 Wt + 5.49
	Females (active) 18-35 years: FFM = 0.666 (Ht^2/R) + 0.164 Wt + 0.217 Xc − 8.78
	Females 30-50 years: FFM = 0.536 (Ht^2/R) + 0.155 Wt + 0.075 Xc + 2.87
	Females 50-70 years: FFM = 0.470 (Ht^2/R) + 0.170 Wt + 0.030 Xc + 5.7
	Males 18-30: FFM = 0.485 (Ht^2/R) + 0.338 Wt + 5.32
	Males 30-50 years: FFM = 0.549 (Ht^2/R) + 0.163 + Wt + 0.092 Xc + 4.51
	Males 50-70 years: FFM = 0.600 (Ht^2/R) + 0.186 Wt + 0.226 Xc − 10.9

FFM = fat-free mass in kilograms; Ht = height in centimeters; R = resistance; Wt = body weight in kilograms; Xc = reactance.

Reprinted by permission from T.G. Lohman, *Advances in Body Composition Assessment* (Champaign, IL: Human Kinetics, 1992).

BIA (3.1 kg [6.8 lb]). The authors concluded that the MW estimates using the two methods recommended by the NCAA for collegiate wrestlers (hydrostatic weighing and skinfolds) were the most accurate and precise.

BIA Method and Hydration

Bartok et al. (50) studied the effect of dehydration on MW assessment as measured by skinfolds, leg-to-leg BIA, and whole-body bioimpedance spectrometry (BIS). Acute thermal dehydration affected all approaches, producing large total errors of about 4 kg (9 lb). In a normally hydrated state, the BIA and BIS approaches produced considerably larger total errors than skinfolds. However, the SEE for the BIS method was much lower than for the BIA method, which indicated that BIS when applied to a specific athletic population may do as well as skinfolds (50).

With the development of a simplified, inexpensive BIS analyzer, BIS may have applications to the successful estimate of MW, whereas BIA may not. Clark et al. (51) also concluded that leg-to-leg BIA, compared to the four-component model, showed a large SEE of 3.5 kg (7.7 lb) and was deemed not to be an acceptable approach for MW. In contrast, the work of Utter and Lambeth (46), comparing multifrequency BIA to skinfolds for estimating MW, found the two approaches similar, using hydrostatic weighing as the criterion method. Hetzler et al. (52), in their study comparing skinfolds to whole-body single-frequency BIA, concluded that the two methods cannot be used interchangeably to determine MW in the same sample.

PRACTICAL INSIGHTS

The combination of skinfolds and BIS may provide a practical way of assessing both hydration level and fatness. Although BIS analyzers are more expensive than single frequency BIA equipment, future developmental research may provide for a less expensive BIS approach enabling both hydration and fatness assessment in all participants.

Summary

MW is the lowest weight that can be maintained indefinitely without adverse effects on health and performance. It has operationally been defined as 5% to 7% body fat in older adolescent boys and men and 12% to 14% body fat in older adolescent girls and women. Professional organizations have recommended that athletes not be allowed to compete at less than their MW to ensure their health and safety. MW is determined by estimating the body composition (fat and fat-free masses) and calculating MW. A number of laboratory and field methods are available for estimating body composition that differ in complexity, expense, and accuracy. In general, laboratory methods are more accurate and have been used as the criterion for validating field methods. Skinfolds or BIA measurements are recommended as the most practical approach for determining the MW of athletes in the field.

Applying Body Composition Methods to Specific Populations

Jennifer W. Bea, PhD
Timothy G. Lohman, PhD
Laurie A. Milliken, PhD, FACSM

LEARNING OBJECTIVES

After completing this chapter, you will be able to do the following:

- Understand and describe how the accuracy of laboratory methods of body composition assessment is affected when applied to various populations

- Understand and describe how the accuracy of field methods of body composition assessment is affected when applied to various populations

- Describe a more accurate alternative approach for assessing body composition in a given population

The application of body composition methods and equations to different populations is an essential aspect of body composition assessment. Although the reference methods of computed tomography (CT), magnetic resonance imaging (MRI), and multicomponent models described in chapter 2 of this book apply well to all populations, laboratory and field methods require testing and adjustment in different populations. There is no need to adjust reference methods or results for racial/ethnic, age, sex, or other demographic differences, because these methods provide for valid estimates under different demographic conditions; they measure directly the factors that often vary between populations.

For laboratory methods using the two-component system, it is essential to determine whether a given formula applies to a given population (e.g., density and proportions of the components of fat-free mass [FFM]) when using densitometry and air displacement plethysmography (ADP). Any deviation in these components from that which is assumed when applied to other populations will result in increased errors in the estimate of percent fat. For field methods, such as skinfolds and bioelectrical impedance analysis (BIA), the cross-validation of equations developed on the population under study is an essential requirement. Use of the wrong formula or equation can result in large systematic errors in body composition estimates, even though an excellent measurement protocol was followed. In this chapter, the ability of both laboratory and field methods to provide accurate assessments of body composition in special populations will be discussed.

Laboratory Methods

The laboratory methods that will be discussed here are hydrodensitometry, total body water (TBW), dual-energy X-ray absorptiometry (DXA), and total body potassium (TBK) counting. Hydrodensitometry has been an accepted method for decades, and therefore, its limitations have been well studied. Another two-component model, ADP, is comparatively new but suffers from the same limitations as hydrodensitometry. Because the hydrodensitometry literature is well established and the two laboratory techniques are based on the same theoretical principles, only hydrodensitometry as it is adjusted to various populations will be discussed. TBW is often used to assess the amount of lean mass and relies on the assumption that water is a constant fraction of lean mass. It is also used as part of multicomponent models and is shown to vary among populations. DXA is a three-component model, which was described in chapter 2, whose components are fat, soft tissue lean, and bone masses. Considerations for variously sized individuals will be discussed. Finally, because potassium is found in lean mass, TBK measures can be used to estimate lean mass. This conversion to lean mass varies among populations. Each of these laboratory techniques will be discussed as it is applied to various populations.

Hydrodensitometry

Hydrodensitometry (underwater weighing) is a method by which body composition can be estimated using the two-component model, described in chapter 2, where the body mass is assumed to be made up of only two components, fat and FFM. Although the two-component model estimates of body composition using hydrodensitometry are reasonably accurate in most people, as described in chapter 3, variability among individuals in the water, protein, and mineral fractions of the FFM and in concentrations of other constituents of the FFM results in greater error among those who do not conform to the assumptions (1). Systematic errors may also occur in some subgroups based on age (2,3), health (4), race/ethnicity (5), gender (5), and weight change (6,7) and in some groups of athletes (8-11) in which the density of the FFM is different than assumed. A summary of differences in FFM density is given for various healthy and clinical populations (5) (table 7.1). In these groups, different constants must be used for accurate estimations of body composition using a two-component model. Constants for the estimation of percent total body fat and FFM density by race/ethnicity, gender, athletic status, and age are provided in table 7.1 to account for nonconformance to the classic Siri (1961) (%Fat = 4.95/Db − 4.50) and Brozek (1963) (%Fat = 4.57/Db − 4.4142) two-component models. Without the use of specific constants in special populations, body fat may be over- or underestimated. For example, the classic Siri and Brozek two-component models overestimate body fat by 2% to 5% in prepubescent children (11).

Using the two-component models, Modlesky et al. (8) and Withers et al. (14) found higher water content in the FFM (table 7.1). In contrast, Silva et al. (12) observed that female adolescent athletes, but not males, the majority of whom were postpubescent, had a lower water fraction and a higher protein fraction of FFM. In another investigation, Silva et al. (12) observed that FFM hydration in elite male judo

Table 7.1 Population-Specific Two-Component Model Conversion Formulas

Population	Age	Gender	%BF	FFBd* (g/cm³)
Ethnicity				
African American	9-17	Female	(5.24/Db) − 4.82	1.088 (0.03931 lb/in.³)
	19-45	Male	(4.86/Db) − 4.39	1.106 (0.03996 lb/in.³)
	24-79	Female	(4.86/Db) − 4.39	1.106 (0.03996 lb/in.³)
American Indian	18-62	Male	(4.97/Db) − 4.52	1.099 (0.039704 lb/in.³)
	18-60	Female	(4.81/Db) − 4.34	1.108 (0.04003 lb/in.³)
Asian (Japanese Native)	18-48	Male	(4.97/Db) − 4.52	1.099 (0.039704 lb/in.³)
		Female	(4.76/Db) − 4.28	1.111 (0.040137 lb/in.³)
	61-78	Male	(4.87/Db) − 4.41	1.105 (0.03992 lb/in.³)
		Female	(4.95/Db) − 4.50	1.100 (0.039740 lb/in.³)
Singaporean (Chinese, Indian, Malay)		Male	(4.94/Db) − 4.48	1.102 (0.03981 lb/in.³)
		Female	(4.84/Db) − 4.37	1.107 (0.03999 lb/in.³)

> continued

Table 7.1 *(continued)*

Population	Age	Gender	%BF	FFBd* (g/cm³)
Ethnicity				
Caucasian (10)	4-5.99	Male	(5.34/Db) − 4.93	1.0826 (0.039111 lb/in.³)
	4-5.99	Female	(5.35/Db) − 4.94	1.0821 (0.03909 lb/in.³)
	6-7.99	Male	(5.24/Db) − 4.83	1.0865 (0.03925 lb/in.³)
	6-7.99	Female	(5.17/Db) − 4.74	1.0899 (0.03938 lb/in.³)
	8-9.99	Male	(5.19/Db) − 4.77	1.0887 (0.03933 lb/in.³)
	8-9.99	Female	(5.15/Db) − 4.72	1.0905 (0.03940 lb/in.³)
	10-11.99	Male	(5.13/Db) − 4.69	1.0917 (0.03944 lb/in.³)
	10-11.99	Female	(5.11/Db) − 4.67	1.0926 (0.03947 lb/in.³)
	12-13.99	Male	(5.13/Db) − 4.70	1.0914 (0.03943 lb/in.³)
	12-13.99	Female	(5.05/Db) − 4.61	1.0951 (0.03956 lb/in.³)
	14-15.99	Male	(5.11/Db) − 4.68	1.0923 (0.03946 lb/in.³)
	14-15.99	Female	(4.96/Db) − 4.51	1.0996 (0.03973 lb/in.³)
	16-17.99	Male	(4.97/Db) − 4.52	1.0992 (0.03971 lb/in.³)
	16-17.99	Female	(4.91/Db) − 4.45	1.1021 (0.03982 lb/in.³)
	18-19.99	Male	(4.96/Db) − 4.51	1.0995 (0.03972 lb/in.³)
	18-19.99	Female	(4.88/Db) − 4.42	1.1034 (0.03986 lb/in.³)
	20-22.99	Male	(4.92/Db) − 4.47	1.1013 (0.03979 lb/in.³)
	20-22.99	Female	(4.88/Db) − 4.42	1.1037 (0.03987 lb/in.³)
Hispanic		Male	NA	NA
	20-40	Female	(4.87/Db) − 4.41	1.105 (0.03992 lb/in.³)
Athletes				
Resistance trained (8)	24 ± 4	Male	(5.21/Db) − 4.78	1.089 (0.03934 lb/in.³)
	35 ± 6	Female	(4.97/Db) − 4.52	1.099 (0.03970 lb/in.³)
Endurance trained	21 ± 2	Male	(5.03/Db) − 4.59	1.097 (0.03963 lb/in.³)
	21 ± 4	Female	(4.95/Db) − 4.50	1.100 (0.03974 lb/in.³)
Judo before training (12)	22.6 ± 2.9	Male	(4.97/Db) − 4.52	1.099 (0.03970 lb/in.³)
Judo after training (12)	22.6 ± 2.9	Male	(4.94/Db) − 4.48	1.102 (0.03981 lb/in.³)

Population	Age	Gender	%BF	FFBd* (g/cm³)
Athletes				
Bodybuilders (13) 12 weeks precompetition	35.3 ± 5.7	Female	(4.99/Db) − 4.55	1.0980 (0.03967 lb/in.³)
6 weeks precompetition	35.3 ± 5.7	Female	(4.97/Db) − 4.53	1.0988 (0.03970 lb/in.³)
3-5 days precompetition	35.3 ± 5.7	Female	(4.94/Db) − 4.48	1.1007 (0.03977 lb/in.³)
Bodybuilders (14) 10 weeks precompetition	26.0 ± 4.8	Male	(5.11/Db) − 4.67	1.0943 (0.03953 lb/in.³)
5 days precompetition	26.3 ± 4.7	Male	(5.08/Db) − 4.64	1.0946 (0.03954 lb/in.³)
All sports	18-22	Male	(5.12/Db) − 4.68	1.093 (0.03949 lb/in.³)
	18-22	Female	(4.97/Db) − 4.52	1.099 (0.03970 lb/in.³)
Anorexia nervosa	15-44	Female	(4.96/Db) − 4.51	1.101 (0.03978 lb/in.³)
Clinical population**				
Cirrhosis (Child A)			(5.33/Db) − 4.91	1.084 (0.03916 lb/in.³)
Child B			(5.48/Db) − 5.08	1.078 (0.03895 lb/in.³)
Child C			(5.69/Db) − 5.32	1.070 (0.03866 lb/in.³)
Obesity	17-62	Female	(4.95/Db) − 4.50	1.100 (0.03974 lb/in.³)
Spinal cord injury (paraplegic/quadriplegic)	18-73	Male	(4.67/Db) − 4.18	1.116 (0.04032 lb/in.³)
	18-73	Female	(4.70/Db) − 4.22	1.114 (0.04025 lb/in.³)

*FFBd = fat-free body density based on average values reported in selected research articles.

**There are insufficient multicomponent model data to estimate the average FFBd of the following clinical populations: coronary artery disease, heart/lung transplants, chronic obstructive pulmonary disease, cystic fibrosis, diabetes mellitus, thyroid disease, HIV/AIDS, cancer, kidney failure (dialysis), multiple sclerosis, and muscular dystrophy.

NA = no data available for this population subgroup; %BF = percent body fat; Db = body density.

Adapted by permission from V. Heyward and D.R. Wagner, "Body Composition Definitions, Classification, and Models," in *Applied Body Composition Assessment*, 2nd ed. (Champaign, IL: Human Kinetics, 2004), 9.

athletes decreased from 72% to 71% from a period of weight maintenance to time before a competition, significantly deviating from the assumed value from mammal studies (73.2%) in both periods of the season. Santos et al. (15) have also verified that basketball players presented a lower FFM hydration during a competitive period. A number of investigators have found variation in FFM density (table 7.1). Ideally, in developing valid equations for estimating body fat, a multicomponent model is needed for many athletic populations.

> ## PRACTICAL INSIGHTS
>
> The traditional two-component model will result in higher errors when populations vary from the assumed fat-free body density of 1.1 g/mL. This value can be adjusted using the values presented in table 7.1. For example, for a Caucasian 10-year-old male, body fatness measured from the two-component model (using either hydrodensitometry or air displacement plethysmography) will likely be overestimated by 3% to 8% unless the density of fat-free mass is adjusted from 1.1 g/mL to 1.0917 g/mL. Percent fat would be calculated as (5.13/body density) minus 4.69. Similar adjustments have been shown to be needed in athletes such as college-aged wrestlers who have a fat-free body density lower than 1.1 g/mL. Refer to table 7.1 for other populations when using the two-component model.

Total Body Water

The TBW method, which is measured by the ingestion and subsequent measurement of an isotope that disperses throughout the water compartment, has long been thought to be unaffected by demographic differences other than age (16) because of the assumption that FFM water content is constant across populations. However, FFM water content has been known to be affected by hydration status; the obese and individuals undergoing weight loss or weight gain may be particularly vulnerable to this type of bias (6,17,18). Variation in the water content of FFM significantly underestimated fat mass (FM) by 0.62 ± 1.56 kg (1.37 ± 3.44 lb), whereas densitometry overestimated it by 0.72 ± 1.6 kg (1.59 ± 3.5 lb) in a study of adults who had lost or gained >3% of initial weight over 23.5 to 43.5 months (6). Obese children have been shown to have greater FFM hydration and reduced density as well, with a slight increase in mineralization according to three- and four-component models (19). Therefore, the assumptions of two-component models, such as hydrodensitometry and TBW, may be violated in obese children and adults. In some conditions, such as pregnancy, the hydration factor may change by approximately 2% (20). However, two-component models are practical and appropriate in this population because other, more accurate methods might be inappropriate because of radiation exposure (e.g., DXA). The technician must weigh the risks and benefits of each method to find the most appropriate method for the given condition.

Dual-Energy X-Ray Absorptiometry

DXA measures three body composition components—fat mass, soft tissue lean mass, and bone mass—and typically provides good estimates of all three across most populations, as described in chapter 3. For this method, several factors affect validity, which include exercise and the consumption of foods and fluids prior to measurement, variations in the size of the person measured (thickness, width, and length), and the software variations available (among manufacturers, updates over time, and for different populations). Researchers are also limited by the proprietary nature of the analytical software because the equations cannot be studied.

When using DXA in many clinical situations or for the measurement of athletes, it may be difficult to control the testing conditions. However, recent evidence suggests that having subjects fast and rest prior to body composition assessment may optimize DXA measurements (21,22), although most laboratories still allow ad libitum food and fluid prior to measurement. Food intake caused measurement errors for trunk fat of 5% to 6% (22), and prior exercise resulted in errors for FM of 10% (21). Best practice is to perform measurements under fasting and rested conditions whenever possible and to perform follow-up measurements under the same conditions as those used in the prior measurement.

Body sizes, such as extreme obesity and tall or broad body size, and weight changes affect the results of DXA (6,18). Given the limited scan area on the DXA table, tall and broad persons often cannot be completely scanned, so the amount of error increases as the proportion of the body that exceeds the dimensions of the table increases. One technique for accommodating broad subjects is to sum left and right body partial scans (23), although tissue thickness may introduce some error (24). Summing partial scans for tall persons resulted in an overestimation of FFM by an average of 3 kg (7 lb) and FM by 1 kg (2 lb) and unacceptable technical errors (23). Therefore, the summation of partial scans using the technique of Nana (23) is an acceptable modification for broad people, but an alternative is still needed for tall people.

Pourhassan (6) showed DXA significantly underestimated the gain in FM (by 1.73 ± 2.25 kg [3.81 ± 4.96 lb]) and overestimated the gain in FFM (by 2.39 ± 2.78 kg [5.27 ± 6.13 lb]) in those who gained weight compared to a model. If precise measures of body composition are desired following weight gain or loss, the four-component model and MRI can be used to provide greater protection against bias. Weight changes often occur in elite athletes during the course of training or competition, so adjustments may need to be made if you are using DXA (5,25,26). Overestimation of FM at the lower ends and underestimation at the upper ends may be present when using DXA (25), which may be caused by alterations in soft tissue hydration. When tracking body fatness in athletes with low levels of body fatness, especially in sports where variations in hydration status are common, DXA should be used with caution.

Lastly, the DXA manufacturers provide various software options for different populations. For example, the pediatric software with correction factors for reduced body size and age should be employed along with the manufacturer's guidelines when measuring children. Also, some manufacturers provide an extended analysis option for those who are thicker in depth when they are lying on the table. Though users cannot overcome the limitation of the proprietary nature of the software, they can follow the recommended options provided by the manufacturer. Finally, when software versions are upgraded, we recommend reanalyzing prior scans for those clients who are being scanned for follow-up visits so that comparisons over time are made using similar software versions.

Total Body Potassium (TBK) Counting

TBK counting is a laboratory method discussed in chapter 2. Naturally occurring radioactive potassium can be counted from humans using detectors designed for that purpose. Because TBK FFM ratios have been identified, TBK counting can be used to determine body composition. However, adjustments may be needed for TBK because TBK per kilogram of FFM increases throughout childhood to reach a constant value in late adolescence. The adult value of 2.66 g TBK/kg FFM is based on four adult cadavers (27). Additional estimates of TBK per FFM for Caucasian and African American male and female adults separately across various age ranges have been summarized elsewhere (28). For children, the TBK to FFM ratio is lower and on the order of 2.11 to 2.31 g TBK/kg FFM for girls and 2.31 to 2.42 g TBK/kg FFM for boys (29,30). When using TBK as a body composition method or as part of a multicomponent method, it's important to select the appropriate TBK to FFM ratio for the population being studied.

Field Methods

Field methods that will be discussed in the following sections are skinfold thicknesses, ultrasound, BIA, and body mass index (BMI). Various populations will be discussed, such as adults and children, older adults, and athletic populations.

Investigator-Specific Versus Population-Specific Equations

For field methods, research suggests that the use of generalized equations such as Jackson–Pollock skinfolds may not apply to specific populations such as children, athletes, or female athletes. As a result, there are hundreds of equations that have been published for estimating body composition in specific populations. In many cases, these equations may be investigation or investigator specific, meaning a given investigation may have used a measurement protocol unique to the study, such as measuring skinfolds using a procedure that was not used by previous authors, or another bias was present in the study design.

Lohman et al. (31) showed the effect that the investigator, the caliper type, and the choice of equation had on body fat estimates. Four experts used four different calipers to measure five skinfold sites using the same written procedures as those used in the Jackson–Pollock equations (32). Finally, five different equations were used to estimate percent fat to show the effect of using the wrong equation. Lohman (31) found considerable differences among investigators in three of the five skinfold sites (percentage differences of 25% to 40% for the thigh, suprailiac, and abdomen sites), showing low objectivity. Also, there was an interaction between caliper and investigator with a large difference in measurements for some investigators but not for all. The choice of equation, caliper, and investigator resulted in body fatness that ranged from 14.1% to 28.1%. To further illustrate the effect of the investigator on skinfold assessment, Lohman (33) compared two studies measuring the same four skinfolds in female athletes and found that, in total skinfolds, there was a 14 mm (0.55 in.), or 3%, fat difference between studies. These studies illustrate the importance of following a standardized measurement technique, using a skinfold caliper the investigator is familiar with, and choosing the most appropriate validated equation for the person being measured. Also, when evaluating research studies, readers should evaluate whether the methods used are standardized methods that are generally applicable and whether the study has been cross-validated by others in the field. The goal is for published equations for any field method to be generalized equations rather than investigator- or investigation-specific equations.

Skinfold Method

In chapter 4, the various approaches to developing a generalized skinfold equation were summarized. One of the most generalized approaches recently published is by Stevens et al. (34), where equations are presented to cover children and adults of all ages using the representative samples of the National Health and Nutrition Examination Survey (NHANES) III and IV (1998-2004).

In nonathletic adults, the most valid equations for body fat are the Jackson–Pollock equations, using the sum of three, four, or seven skinfolds (35,36). The research protocol for measuring three, four, or seven skinfolds is well described in the Jackson and Pollock study (37). Despite their widespread use, Peterson et al. (38), using the four-component model, concluded that the Jackson–Pollock equations overestimate percent fat by 6% in adult women. This unexpected result can be explained by a number of discrepancies in the Peterson et al. approach. These authors used the wrong measurement protocol for skinfolds, reporting unusually large standard errors of estimate (SEEs) for the new equations developed and did not ensure the four-component model was accurate by reporting the water as a fraction of fat free mass (W/FFM) variation as recommended by Lohman et al. (39). They also left out the statistical procedures leading to the four skinfolds recommended, did not cite the exact Jackson–Pollock equations being evaluated, and did not modify the Bland–Altman procedures as recommended by Hopkins (40) when using a prediction equation (X) as compared to a reference method. For these reasons, the Peterson approach is not recommended. More recently, Jackson et al. (41) has evaluated the Jackson–Pollock equations using DXA as the criterion method. These authors found that the Jackson–Pollock equations could be modified to better apply to socially and ethnically diverse young men and women (table 7.2). However, there is evidence that DXA underestimates FM and may limit the accuracy of the recommended equations (42).

In children and youth, we recommend the following Lohman equations (33) with either triceps and calf skinfolds or triceps and subscapular skinfolds. Successful cross-validation of these equations has been carried out by Roemmich et al. (2) and Wong et al. (43) using the four-component model. It is important to note that the use of the triceps plus subscapular skinfold quadratic equation does not apply for subjects with high skinfold measurements, and it is essential to use the linear equations when the sum of the skinfolds (ΣSF) > 35 mm (1.38 in.).

Table 7.2   Traditional Jackson–Pollock Equations and Recommended Revised Skinfold Equations for Young White, Hispanic, and African American Females and Males

		R	SEE
Jackson–Pollock two-component Siri equations (37)			
Females	%BF-GEN (0.4453 × ΣF) − (0.0010 × ΣF²) − 0.5529	0.84	3.9
Males	%BF-GEN (0.3460 × ΣM) − (0.0006 × ΣM²) − 3.9428	0.91	3.4
DXA equations from Jackson et al. (41)			
White females	%BF-DXA = (0.4446 × ΣF) − (0.0012 × ΣF²) + 4.3387	0.86	3.8
Hispanic females	%BF-DXA (0.4446 × ΣM) − (0.0012 × ΣF²) + 6.7066	0.83	3.4
White males	%BF-DXA = (0.2568 × ΣF) − (0.0004 × ΣF²) + 4.8647	0.92	3.0
Hispanic males	%BF-DXA = (0.2568 × ΣF) − (0.0004 × ΣF²) + 5.5458	0.91	3.0
African American males	%BF-DXA = (0.2568 × ΣF) − (0.0004 × ΣF²) + 3.8954	0.95	2.6

SEE = standard error of estimate; %BF-GEN = body fat percentage from Siri's two-component percentage fat equations; %BF-DXA = body fat percentage from DXA; ΣF = sum of triceps, suprailium, and thigh skinfolds; ΣM = sum of chest, abdomen, and thigh skinfolds using Jackson et al. (37,41) skinfold protocol.

Triceps and Calf Skinfolds

Males, all ages (7-17 years):

$$\%Fat = 0.735\ \Sigma SF + 1.0 \tag{7.1}$$

Females, all ages (7-17 years):

$$\%Fat = 0.610\ \Sigma SF + 5.0 \tag{7.2}$$

Triceps and Subscapular Skinfolds (>35 mm [>1.38 in.])

Males (where I is the intercept from table 7.3):

$$\%Fat = 0.783\ \Sigma SF + I \tag{7.3}$$

Females:

$$\%Fat = 0.546\ \Sigma SF + 9.7 \tag{7.4}$$

Triceps and Subscapular Skinfolds (<35 mm [<1.38 in.])

Males (where I is the intercept from table 7.3):

$$\%Fat = 1.21\ (\Sigma SF) - 0.008\ (\Sigma SF)^2 + I \tag{7.5}$$

Females:

$$\%Fat = 1.33\ (\Sigma SF) - 0.013\ (\Sigma SF)^2 + 2.5\ (2.0\ Blacks,\ 3.0\ Whites) \tag{7.6}$$

Because the intercept varies with maturation level and racial group for males, use the intercepts given in table 7.3 in place of I in equations 7.3 and 7.5.

Thus, for a White pubescent male with a triceps of 15 and a subscapular of 12, the percent fat would be

$$\%Fat = 1.21\ (27) - 0.008\ (27)^2 - 3.4 = 23.4\%$$

Table 7.3 Intercepts for Males in Equations 7.3 and 7.5*

Age	Black	White
Prepubescent	−3.5	−1.7
Pubescent	−5.2	−3.4
Postpubescent	−6.8	−5.5
Adult	−6.8	−5.5

*Calculations were derived using Slaughter et al.'s (44) equation.

When $\Sigma SF > 35$ mm (1.38 in.), such as if the sum of the triceps and subscapular is 40 mm (1.57 in.), then

$$\%Fat = 0.783 \,(40) - 3.4 = 27.9\% \tag{7.7}$$

Stevens et al. (45) published prediction equations for children 8 to 17 years using the 1999 to 2003 NHANES national probability data with adjustments for Hispanic and Black children with DXA as the criterion method. Because DXA is not a reference method, it would be important to cross-validate the Stevens et al. equations using the four-component model. Until that time, the prior equations (7.1-7.6) are recommended.

PRACTICAL INSIGHTS

Skinfold equations that have been validated against a four-component model are accurate within 3% to 4% body fatness when the technician is trained and uses the correct protocol. This level of accuracy is very good for a field method. These selected skinfold equations are for children and athletes. The key to getting this level of accuracy is following the protocol established for the equation being used and being properly trained in the skinfold technique. It is best to get face-to-face training with a professional to be sure you are conducting measurements properly. Becoming certified by a reputable organization is recommended.

In athletes, we have the work of Evans et al. (46), using skinfolds and the four-component model, with a general formula of three (triceps, abdomen, and thigh) or seven (tricep, subscapular, abdomen, thigh, chest, midaxillary, and suprailiac) skinfolds and an adjustment for sex and race.

$$\Sigma 3 \text{ skinfolds} = \%Fat = 8.997 + 0.247 \,(3SK) - 6.343 \,(sex) - 1.998 \,(race) \tag{7.8}$$

$$\Sigma 7 \text{ skinfolds} = \%Fat = 10.566 + 0.121 \,(7SK) - 8.057 \,(sex) - 2.595 \,(race) \tag{7.9}$$

where sex = 0 for female and 1 for males and race = 0 for Whites and 1 for Blacks.

There is a lack of research in older adults despite the fact that skinfold compressibility changes with advancing age. In older adults, the study performed using a multicomponent model is by Williams et al. (47), who developed a skinfold prediction equation.

$$\text{Men \%Fat} = 0.486 \, (\Sigma 4SK)_m - 0.0015 \, (\Sigma 4SK)^2 + 0.67 \, \text{Age} - 3.83 \qquad \textbf{(7.10)}$$

$$\text{Women \%Fat} = 0.488 \, (\Sigma 4SK)_f - 0.0011 \, (\Sigma 4SK)^2 + 0.127 \, \text{Age} - 3.01 \qquad \textbf{(7.11)}$$

where

$$\Sigma 4SK_m = \text{chest} + \text{subscapular} + \text{midaxillary} + \text{thigh} \qquad \textbf{(7.12)}$$

and

$$\Sigma 4SK_f = \text{triceps} + \text{subscapular} + \text{abdomen} + \text{calf} \qquad \textbf{(7.13)}$$

Ultrasound

Ultrasound technology can be applied to the measurement of subcutaneous fat layers, which are related to total body fat in ways similar to the relationship between skinfold thickness and total body fat. However, the relationship between ultrasound fat and percent fat by a reference method may differ among different populations, as is the case for many other methods discussed in this chapter. For the past 30 years, ultrasound has proven to be similar to skinfolds in accurately estimating percent fat (chapter 3). However, the recent approach of Müller et al. (48,49,51) and Störchle et al. (50), using a semiautomatic approach, allows for more reliable and accurate estimates of fat thickness among observers. Intraobserver reliability improved from an intraclass correlation coefficient (ICC) of 0.968 and a SEE of 0.60 mm (0.24 in.) (48) to an ICC of 0.998 and a SEE of 0.55 mm (0.22 in.) (51). Thus, future use of ultrasound may be warranted where accurate measures of subcutaneous fat are needed. Also, ultrasound may provide a more valid measure of percent fat, especially in the lean athletic population, because it has the sensitivity to detect small changes over time without compression of the fat layer (51). Future research comparing ultrasound and skinfolds to percent fat using the four-component model will establish the usefulness of this new approach in different populations.

Bioelectrical Impedance Analysis

In chapter 4, different approaches to the measurement of BIA were reviewed and, because there are very different hardware as well as measurement configurations, it is important to revise equations that were developed using different hardware. First, the single-frequency arm-to-arm and leg-to-leg approaches will be discussed; then the whole body single frequency for four electrodes; and finally, the multifrequency with eight electrodes. Each section will be followed by a discussion of recommended equations for various populations. Although many BIA equations have been developed on specific populations of all ages, such as those described by Heyward and Wagner (52) and Mialich et al. (53), many of these equations have not been cross-validated using the four-component method and may have an investigation bias. Equations that are recommended are those that have been tested against a four-component model, have acceptable rates of errors and low bias, and have been cross-validated.

Single-Frequency BIA

In general, the arm-to-arm and leg-to-leg approaches to estimating body composition are less accurate than total body BIA (arm-to-leg approach). There are both commercial and laboratory hardware that are designed for measuring arm or leg resistance. Leg resistance has been carefully studied by comparing BIA to four-component models in the work of Bartok et al. (54) and Clark et al. (55) in groups of college wrestlers. Minimum weight is an important measurement in collegiate wrestling because wrestlers need to maintain a certain minimum weight and at least a 5% body fat level to be eligible for competition. Clark et al. (55) compared the leg-to-leg BIA to the four-component model in 57 college-aged male wrestlers. Minimum weight was predicted with a SEE of 3.4 kg (7.5 lb), and the error was significantly correlated with subject weight and percent fat, showing bias compared to the four-component model. The prediction of body water by the leg-to-leg BIA system also yielded unacceptably large SEEs, and the authors concluded that the leg-to-leg approach was not an acceptable method for predicting minimum weight in wrestlers. Additionally, Bartok et al. (54) used the leg-to-leg BIA approach and skinfolds with the four-component model in dehydrated and euhydrated wrestlers. In the euhydrated subjects, skinfolds predicted the minimum weight the best (SEE = 2.0 kg [4.4 lb]), and the SEE for leg BIA was 3.5 kg (7.7 lb). The SEEs were higher in the dehydrated wrestlers for all methods. Earlier research by Jebb et al. (56) and Nunez et al. (57) indicated that the leg-to-leg BIA system was less accurate than the whole-body approach by a small degree. However, the SEEs were larger than usual in both studies, and the body composition estimates were not much improved over BMI (bias ± 95% limits of agreement for BIA and BMI vs. the four-component model were 0.9 ± 10.2 and −0.9 ± 10.8, respectively) (56). Based on this review, whole-body BIA is a more accurate body composition assessment compared to an arm-to-arm or a leg-to-leg approach for general use or research.

PRACTICAL INSIGHTS

Equations for estimating body water and body fatness using single-frequency BIA as a field method require careful standardization of subject hydration and exercise levels. Therefore, it is important to follow all pretest precautions when performing measurements when using the single-frequency BIA method, especially those associated with ensuring proper hydration. Chapter 4 gives these precautions in detail. A client should not be measured if they arrive for testing without following the precautions because the error of measurement will be inflated when the hydration level is not normal. In addition, it is helpful when remeasuring a client over time to follow a standardized protocol under similar conditions so that hydration shows less variation from one time point to the next.

Other applications of BIA include phase angle analysis and bioelectrical impedance vector analysis. A review of these approaches is given by Bera (58). An example of the application of BIA using impedance vector analysis in patients with chronic obstructive pulmonary disease is described by Walter-Kroker et al. (59) and may provide useful tools for determining fluid status, body cell mass, and extracellular cell mass. Though these areas are beyond the scope of this book, they may be of interest to the reader.

BIA Equations for Adults BIA equations in adults for estimating FFM are given in table 7.4. These equations have been recommended by Moon (60) for athletes. However, Moon cautions their use because generalized athlete-specific BIA equations have not yet been developed using multicomponent models. The same can be said about using them in adults because, at this time, different investigators obtain different results using BIA. Elia (61) observed the SEE for predicting body water in adult men and women to be large (2.5 kg [5.5 lb] women and 3.5 kg [7.7 lb] men), with similar findings with weight and height. For Kushner and Schoeller (62), the SEEs are much smaller, as shown in chapter 4. Part of the differences among investigations may be the heterogeneity of the sample under study, with more heterogeneity resulting in larger SEEs. Thus, the equations in table 7.4 are recommended for BIA.

Additional equations for TBW and FFM based on the four-component model are presented in table 7.5 as they have been used to generate national norms using data from NHANES III (63).

Table 7.4 Recommended FFM Equations for BIA in Adults

Sex and age groups	FFM equations
Females 18-30 years	FFM (kg) = 0.476 (Ht^2/R) + 0.295 Wt + 5.49
Females (active) 18-35 years	FFM (kg) = 0.666 (Ht^2/R) + 0.164 Wt + 0.217 Xc − 8.78
Females 30-50 years	FFM (kg) = 0.536 (Ht^2/R) + 0.155 Wt + 0.075 Xc + 2.87
Females 50-70 years	FFM (kg) = 0.470 (Ht^2/R) + 0.170 Wt + 0.030 Xc + 5.7
Males 18-30 years	FFM (kg) = 0.485 (Ht^2/R) + 0.338 Wt + 5.32
Males 30-50 years	FFM (kg) = 0.549 (Ht^2/R) + 0.163 Wt + 0.092 Xc + 4.51
Males 50-70 years	FFM (kg) = 0.600 (Ht^2/R) + 0.186 Wt + 0.226 Xc − 10.9

Ht = height in centimeters; Wt = body weight in kilograms; R = resistance (ohms); Xc = reactance (ohms).

Table 7.5 Recommended TBW and FFM Equations Using BIA and NHANES III Participants

Males	TBW = 1.203 + 0.176 Wt + 0.449 Ht^2/R	r^2 = 0.84, RMSE = 3.8 L
Females	TBW = 3.747 + 0.113 Wt + 0.45 Ht^2/R	r^2 = 0.79, RMSE = 2.6 L
Males	FFM = −10.678 + 0.262 Wt + 0.652 Ht^2/R + 0.015 R	r^2 = 0.90, RMSE = 3.9 kg
Females	FFM = −9.529 + 0.168 Wt + 0.696 Ht^2/R + 0.016 R	r^2 = 0.83, RMSE = 2.9 kg

Wt = weight in kilograms; Ht = height in centimeters; R = resistance in ohms; RMSE = root mean square error.

BIA Equations for Children and Youth   There are concerns with accurately assessing body fat and FFM in children and youth because of age, maturity, and changes in body shape with growth and development that may change the relationship between resistance index and body composition. Montagnese and colleagues (64) have addressed the question of applying a single BIA equation across the ages of 4 to 24 years, as Kushner et al. (65) have done. In a large sample of children and young adults, Montagnese et al. (64) found that age or pubertal status affected the relationship between resistance index and FFM, decreasing the SEE from 2.8 kg (6.2 lb) (without age effects) to 2.6 kg (5.7 lb). Kushner et al. (65) studied 37 prepubertal and 44 preschool children and found that significant bias existed in only the preschool children when BIA was compared to TBW. Thus, it would appear that equations for children and youth need to be developed separately to improve the FFM prediction.

The equations of Houtkooper et al. (66) are based on a sample of ninety-four 8- to 15-year-old boys and girls. This Houtkooper et al. equation was cross-validated by Wells et al. (67) using the four-component model. Although the Montagnese et al. study (64) predicted FFM with a SEE of 2.9 kg (6.4 lb), the Houtkooper et al. study (66) predicted FFM with a SEE of 2.1 kg (4.6 lb).

$$FFM = 0.61\ Ht^2/R + 0.25\ Wt + 1.31 \tag{7.14}$$

For male and female children who are athletes, Moon (60) recommends Lohman's (33) equation.

$$8\text{-}15\text{ years: FFM (kg)} = 0.620\ (Ht^2/R) + 0.210\ Wt + 0.100\ Xc + 4.2 \tag{7.15}$$

Multifrequency BIA Methods

Multifrequency devices are typically not commonly available for public use and are largely used in research and medical facilities. The advantage of these devices is that a higher frequency allows the current to penetrate into cell membranes and can measure intracellular cell volumes in addition to the traditional body composition measures (57). The best results for BIA in estimating body composition are from using eight-electrode, multifrequency BIA from the work of Bosy-Westphal et al. (68) (chapter 4). With corrections for trunk index, both FFM and body water can be predicted better than from single-frequency BIA.

Body Mass Index

BMI, which was discussed in chapter 4, is an easy height for weight measurement that is particularly helpful when measuring large populations. It is typically used as a proxy measure for overweight and obesity where it is assumed that more weight is from the presence of more fat. This assumption is violated in some populations, especially those who are overly muscular. But there are also concerns about the relationship between BMI, obesity, and disease risk and how this varies among populations. These issues will be discussed in the following sections, first in adults and then in children.

BMI in Adults

Researchers have questioned whether BMI can distinguish between those with greater adipose or body fat equally well across various age, sex, and racial/ethnic backgrounds. BMI as a surrogate estimate of adiposity is age and sex dependent among adults (69). In one study, BMI was found to be independent of race/ethnicity after adjusting for sex and age, though the study was limited to 202 Black and 504 White adults (1,69). However, several other studies have demonstrated racial and ethnic differences in the relationship between BMI and total body fat (TBF). A comparison of European, Maori, Pacific Island, and Asian Indians found significant differences between BMI and body fat among the different races/ethnicities (70). For example, a fixed 29% TBF for men and 43% for women was equivalent to a BMI of 30 kg/m² in Europeans, but approximately 5 BMI units higher for Pacific Islanders and 5 units lower for Asian Indians. If BMI were fixed at 30 kg/m², the corresponding percent TBF was 9% to 12% lower in Pacific Islanders than in Asian Indians (TBF 25% vs. 37% in men, respectively, and 38% and 47% in women, respectively) (70).

An earlier review comparing Asians as a group (Indonesians [Malays and Chinese ancestry], Singaporean Chinese, Malays and Indians, and Hong Kong Chinese) to Caucasians observed that Asians demonstrated higher body fat for a given BMI (3%-5%) but also noted that there is variability within the group of Asians (71). In another study, the relationship between BMI and percent TBF was compared in a sample of 487 men and 933 women from various backgrounds (Hispanic Americans, African Americans, and European Americans) by Fernandez et al. (72). The samples of men and women were 30.2% and 19.4% Hispanic, 30.4% and 32.6% African American, and 39.4% and 48.0% European American, respectively. There were no significant differences in the relationship between BMI and percent TBF between men of Hispanic and European, African American and European, or Hispanic and African American backgrounds. However, there were significant differences in percent TBF predicted by BMI between Hispanic and European (P < 0.002) and African American and Hispanic (P = 0.020) women (but not African American and European American women). Further, a BMI < 30 was associated with more body fat in Hispanics than European Americans and African Americans, but a BMI > 35 was associated with greater body fat among European Americans than African Americans or Hispanics (72). As a representative sample, these studies indicate that racial and ethnic differences do exist in the relationship between BMI and TBF, which suggests BMI may be an unsuitable proxy measurement for TBF.

The BMI cut points related to morbidity and mortality were created in samples with predominantly Caucasian participation. Even among Caucasians, BMI has been shown to be highly specific but not sensitive in terms of obesity diagnosis. Specificity is the ability of a test (in this case BMI) to correctly identify someone as being disease free (not having excess adiposity), and sensitivity is the ability of a test (BMI) to correctly classify someone as having a disease (excess adiposity). A meta-analysis of 32 studies found specificity to be 90%, whereas sensitivity was 50% overall. However, some differences were seen across the studies. The Chinese demonstrated much higher sensitivity in general (>80% for five of seven studies) compared to those conducted in the United States (range: 43%-69% across 13 studies), whereas other

Asian countries demonstrated lower sensitivity to diagnose obesity from BMI (India, Singapore, Thailand, Japan; range: 8.9%-34% across five studies) (73).

Studies conducted in the United States include various racial and ethnic groups, but they tend to be weighted toward Caucasians. Three of the 13 U.S. studies demonstrated lower specificity, but overall, specificity was high across all countries of origin (73). The low sensitivity may translate into underdiagnosing excess adiposity in many groups because of a lower lean and bone mass for a given height, and those at risk for higher morbidity and mortality from excess adiposity may not receive appropriate lifestyle counseling in the clinic or community.

Although BMI has been used indiscriminately among adults for many years, research indicates that it may also have limited sensitivity and specificity to detect adiposity among older adults as well. Older adults tend to experience a shift in body composition, the loss of lean mass and bone mass, and a gain of fat mass as they continue to age, even if they remain weight stable. Without significant loss of height, this phenomenon would render BMI stable. Additionally, as adults approach old age, the prevalence of chronic diseases increases. Chronic disease has been shown to alter body composition, independent of weight, thereby reducing the ability for BMI to adequately estimate adiposity (72,74,75).

In summary, universal BMI cut points are not appropriate for the adult population. The association between racial/ethnic-specific cut points, based on the respective relationships with body fat, needs additional exploration in terms of morbidity and mortality outcomes as compared to other, more sophisticated methods.

PRACTICAL INSIGHTS

BMI in adults has a large SEE of 5% to 6% when used to estimate body fatness. Thus, many individuals with a BMI between 25 and 30 may not be overfat even though they are classified as overweight. Others may be obese with a fat level greater than 32% even though their BMI is below 30. Conversely, there may be muscular people who appear to be obese with a BMI value greater than 30, but their percent fat could be low due to their high muscularity. Whenever possible, it would be more beneficial to measure body fatness to determine health rather than BMI because BMI cannot distinguish between fat and lean masses. Large scale epidemiological studies rely on BMI measurements, which limit conclusions about body composition.

BMI in Children

Child and adolescent overweight and obesity are typically defined by age- and gender-specific percentiles of BMI (i.e., normative data) (www.cdc.gov/growthcharts). The approach is convenient and cost effective for large-scale screening and may be particularly useful for tracking. For example, the predictive value of a child's BMI for being overweight at 35 years increases from childhood to early adolescence and from early adolescence to later adolescence, especially for those in the 95th percentile during childhood (11). However, BMI leaves much to be desired in terms of estimating body composition.

BMI as an estimate of fatness in children has shown a large prediction error for the individual (5%-7% SEE). Those above the 95th percentile of BMI for their age are more likely to carry excess adipose; however, BMI is less discriminating of excess adipose among children between the 85th and 95th percentile from variations in maturity, muscle mass, leg length, and bone mass. Additionally, those with greater than average FFM for height, such as those who might be athletic with more muscle mass, are also susceptible to misclassification (11). A recent systematic review ($N = 37$ articles) and meta-analysis ($N = 33$ articles) by Javed et al. (76) in children ranging from 4 to 18 years of age ($N = 53,521$ patients) confirmed that BMI had high specificity but limited sensitivity to detect excess adiposity. The pooled sensitivity to detect high adiposity was 0.73, and specificity was 0.93. The diagnostic odds ratio was 37 (95% confidence interval: 21-66). Among males, the pooled sensitivity was 0.67, and pooled specificity was 0.94; for females, the pooled sensitivity was 0.71, and the pooled specificity was 0.95. This means that BMI failed to identify over 25% of children with excess body fat. Differences between races explain some of the variance, as did differences in study BMI cut points, reference criteria, and reference methods to assess adiposity.

An effort to shift BMI norm–based classifications to cut points that have been linked to cardiometabolic risk factors and aligned with gender-specific growth curves is underway to address some of the current limitations of the BMI norm–based classification system. The new cut points were developed with the use of receiver operating characteristic (ROC) curve analyses to quantify the sensitivity and specificity of various BMI cut points for percent fat standards such that new BMI cut points categorize youth into the same risk categories as do percent fat categories (11,77-79). Utilizing data from the NHANES III (1998-1994), Going et al. (80) demonstrated a strong relationship between percent fat and cardiovascular disease risk factors among children aged 6 to 18 years, particularly among boys with 20% and girls with 30% TBF. The interaction between age (but not race) and percent body fat was a significant predictor of these risks. Because of the broad adoption of FitnessGram by several professional groups, which includes the new standards, as does the National Youth Fitness Test promoted by the President's Council on Youth Fitness and Sports (11), widespread use of BMI is occurring in U.S. schools. Further validation of the new standards in other populations is needed.

Obesity is not the only concern in children and adolescents that is monitored by BMI. BMI-for-age percentiles are also used by clinicians to screen for malnutrition and identify eating disorders in girls and boys based on the growth of the child compared against the reference population. Athletes, particularly those involved in aesthetic sports, are at high risk of developing eating disorders (11,81). Because of the limitations of BMI during periods of growth, additional clinical markers may be needed to confirm malnutrition and identify eating disorders. Clinicians should understand the potential limitations of BMI when monitoring underweight and subsequent weight gain. Additional measures of body composition should be added so that the composition of weight added is known.

Summary

In this chapter, laboratory and field methods of body composition assessment were reviewed as they apply to various populations. In general, methods should be used where the assumptions of those methods are not violated in the populations that are being measured. For field methods, it is important to choose the method that is the most appropriate given the limitations of the client and then the equation that is appropriate for the demographic characteristics of that individual. Finally, the most appropriate equation is one that has been validated against a multicomponent model in a large sample of people and cross-validated by another research group. The results of these studies should show good accuracy of the equation with low errors compared to the criterion method and minimal bias. These are the equations we have presented in this chapter. Whenever making measurements, always be sure to follow the procedures that were performed in the original validation study.

8

Body Composition Applications

Vanessa Risoul-Salas, MSc, RD
Alba Reguant-Closa, MS, RD
Luis B. Sardinha, PhD
Margaret Harris, PhD
Timothy G. Lohman, PhD
Nuwanee Kirihennedige, MS, RD
Nanna Lucia Meyer, PhD, FACSM

LEARNING OBJECTIVES

After completing this chapter, you will be able to do the following:

- Describe applications of body composition methods to assess nutritional status, growth and aging, sport and exercise, and weight loss and medicine

- Describe advantages of different body composition methods when applied to different fields of study

Body composition methods apply to multiple areas, and although they can be useful in assessing individual and group data, for each application there are methodological as well as practical concerns. This chapter summarizes body composition methodology as it relates to nutritional status of individuals and groups, associations with competitive sport and exercise training, body composition measures during weight loss, body composition issues while recovering from eating disorders, and methodological issues with body composition measures during growth and development, aging, and chronic disease. This chapter was a collaboration among several authors with expertise from several areas.

Nutritional Status

In humans, there is a need to quantify body size and body composition to interrelate an array of influences affecting the organism (1). It is well known that some body composition components, such as muscle and bone mineral content and density, are the result of a combination of genetic, physical activity, dietary, and hormonal factors in both young and older adult populations (2). In contrast, the relation between diet and obesity and the extent to which adiposity represents nutritional status, although widely studied, remains a subject of debate (3).

The assessment of nutritional status includes a comprehensive dietary assessment through a registered dietitian. According to the Academy of Nutrition and Dietetics, this assessment comprises anthropometric (including body composition), biochemical, clinical, dietary, and environmental aspects. Anthropometric, biochemical, and clinical assessments are most useful in combination. For example, it is better to evaluate a patient's body composition in conjunction with a fasting blood glucose test or assessment of blood lipids than assessing body composition alone. In fact, a blood test will provide information about the patient's nutritional status by knowing glucose metabolism or blood lipid levels, such as triglycerides or cholesterol. A blood test in an overweight or underweight patient could also point toward issues with metabolism, such as a thyroid disorder; menopausal changes in estradiol; or issues with liver and kidney function. Of great interest is the recent research on visceral adiposity and its relationship with the metabolic syndrome for predicting insulin sensitivity, glucose tolerance, and dyslipidemia (4). Thus, high body fatness with fat distribution around the waist and torso raises the risk of insulin resistance, abnormal lipids, and inflammation and thus chronic disease.

It is possible for individuals to carry excess fat, including visceral fat, but through an increase in aerobic fitness, their risk of chronic disease may decline. A familiar concept may be Health at Every Size, which was proposed in the 1990s and is based on the underlying causal factors of obesity when leading a sedentary lifestyle with poor nutrition, weight cycling, or other adverse lifestyle habits, as opposed to solely adiposity levels (5). Therefore, it is best to combine body composition assessment with blood parameters because fatness or excessive leanness does not always indicate disease risk or good health because not all obese individuals display a clustering of metabolic and cardiovascular risk factors.

There is an obesity subtype that refers to metabolic health. The first subset, metabolically healthy but obese (MHO), refers to people who, despite having a large amount of fat mass compared with at-risk obese individuals, show a normal metabolic profile but remarkably normal to high levels of insulin sensitivity. A second subset of people, termed the metabolically obese but normal weight (MONW), present normal body mass index (BMI) but have significant risk factors for diabetes, metabolic syndrome, and cardiovascular disease (6).

Especially in overweight and obese individuals, the choice of body composition methodology is critical. Body composition testing in this situation is best done using dual-energy X-ray absorptiometry (DXA) because of the lower invasiveness or discomfort involved compared with densitometry from hydrostatic weighing and air displacement plethysmography (ADP) and lower variability compared to using skinfolds. DXA also tracks lean bone mineral during weight loss to give better estimates of the composition of weight loss. Newer techniques, such as ultrasound in the obese (7), may yield more accuracy, or if it is not available, simple field methods are also informative (e.g., waist-to-hip ratio and waist circumference).

The assessment of other clinical factors, such as a patient's medical history, can provide further insight regarding health status, and if body composition data are available, this risk can be evaluated more robustly. Further, the use of medication or dietary supplements can influence metabolism, body composition, and nutritional status, as can dietary intake, eating patterns, and eating behavior. Additional body composition data may also be put in context with the client's environmental factors. This may include factors such as temperature, humidity, and altitude, but it may also relate to the environmental factors surrounding the client (e.g., eating environment, food deserts vs. urban garden access), work schedules, or the energy expended in exercise and leisure.

Simply put, a dietary assessment will provide the health professional with more information than body composition testing alone, and the environmental assessment will assist in understanding external factors that might affect overall nutritional status. Thus, measuring and integrating various parameters along with body composition assessment will provide a more valid and reliable diagnosis compared to when either occurs alone.

Body Composition Before Birth

Lean and fat masses begin to form in the intrauterine environment and are highly related to maternal nutrient intake. Nutrient restriction during pregnancy will reduce nutrient supply of fetal myofiber number, influencing the amount and even type of muscle fiber that a person will maintain. Research indicates that maternal dietary restriction results in fiber-type transition to offspring, generally favoring increased type I fiber expression (8). It is also important to note that reduced fetal skeletal muscle growth is not fully compensated for after birth because individuals who are born with low birth weight have lower muscle mass into adulthood (8). Moreover, larger birth weight and greater postnatal growth in the first 1 to 2 years of life have been associated with a higher lean body mass later in life (9). In addition, higher postnatal growth is associated with higher levels of obesity for the 2- and 3-year-old child (10-12). Thus, optimal prenatal nutrition will affect body composition and adult risk of obesity. Low birth weight (<2.5 kg [<5.5 lb]) is associated with lower lean body mass, fast catch-up growth, and higher risk of central obesity in adults (13). For this reason, adequate energy, protein, and micronutrient intake before and during pregnancy protects against adult obesity of the offspring (13). This means that women of child-bearing age who want to become pregnant or already are should be assessed and counseled carefully so that weight gain is gradual, sufficient, and not excessive. Nutritional interventions to reduce the risk of excessive body fatness is critical during these years, and monitoring indexes of fatness (e.g., BMI, waist-to-hip ratio, skinfolds, bioelectrical impedance analysis [BIA], or ultrasound) in women is recommended. Clearly, DXA scans to assess body composition during pregnancy are contraindicated because of radiation exposure and possible harm to the fetus (14).

Diet and Body Composition

The law of thermodynamics governs energy balance in the human body. If a person consumes less energy in the form of food calories than is expended, a negative energy balance will produce a reduction in body weight. Conversely, if more energy is ingested compared to energy expended, a positive energy balance will result in body weight gain. Long-term fluctuations in body weight are usually reflected by changes in body fat stores (15). Although simplified, these are general concepts that help with the understanding that changes in body composition are accompanied by changes in energy balance.

In the modern Western diet, the high consumption of refined sugars and vegetable oils has displaced more nutrient-dense foods (16). As a result, there is a decrease in the nutritional value of food, and although diets may sometimes lack protein (17), they are more often devoid of essential nutrients, such as vitamins, minerals, phytochemicals, and fiber, which increases the risk of nutritional deficiencies in the general population. Nutritional deficiencies are also present in overweight or obese individuals, despite a high caloric intake, which leads to a new concept called "high-calorie malnutrition," a term that describes a state of excess caloric intake occurring in concert with a nutritional deficiency (18). This particular issue appears to be present in low-income areas and underdeveloped countries, where populations are experiencing the nutrition transition and gaining better access to food but, unfortunately, in the presence of Westernized eating patterns that are high in sugar, fat, and salt. Higher body fatness is also associated with vitamin D deficiency because of the sequestration of vitamin D by the large body fat pool (19). Thus, body composition data can provide meaningful insight regarding a person's nutritional risk.

Growth and Development

Measuring body composition in childhood and youth provides for an estimation of risk for chronic disease. Is there a cut point in terms of body fatness (25% boys, 32% girls) (20) associated with an increased risk of higher lipid levels, blood pressure, and fasting insulin and blood glucose levels? How is body composition best assessed to identify this risk (21)? For over 40 years, the Bogalusa Heart Study has been asking these questions in studying the relation of BMI and skinfolds to increased risk of chronic disease. The research of Freedman and Sherry (22) and Freedman et al. (23) demonstrates that skinfolds (triceps), body fatness (skinfold equations), DXA fat estimates, and high levels of BMI at a given age estimate risk levels in children and youth of all ages. Freedman et al. uses percentiles of BMI to assess risk, thus allowing for changes in the relationship during growth and development. These authors found that risk factors are likely to be elevated for chronic disease at a BMI at or above the 95th percentile for age and gender. Going et al. (24), using the National Health and Nutrition Examination Survey (NHANES) III and IV for boys and girls 6 to 18 years of age, found 20% fat for boys and 30% fat for girls (based on skinfolds and the equation of Williams et al. [20]) to be predictors of chronic disease risk.

PRACTICAL INSIGHTS

While the use of BMI is generally not recommended in assessing an individual's body composition, especially for children or athletes, research in children and youth has shown that both BMI and skinfolds are useful in assessing cardiovascular risk. Thus, at present, BMI is widely used in assessing health-related fitness in children throughout the country in educational settings. Properly presented information on health risks for a given child can be an important prevention approach to a lifetime risk of overfatness. Both the President's Council on Sports, Fitness and Nutrition and the FitnessGram have combined efforts in this direction.

Central fat and visceral fat have also been linked to increased risk of chronic disease, especially in obese children (25). Key questions for children and youth relate to the use of one cut-point value (e.g., 20% fat for boys and 30% fat for girls) for boys and one for girls across all ages 9 to 16 years. Second, does waist and hip circumference add to or substitute for skinfolds when accounting for disease risk? Third, what are the cut points for waist circumference at difference ages for boys and girls? Many of these questions can be answered from data collected in the past decade by NHANES in children and youth.

In the past, healthy growth in children has been defined by height and weight along with BMI using the national norms to assess an individual's growth. The assessment of muscle and bone growth for clinical applications is not yet well established. Recent developments with bone growth using the peripheral quantitative computer tomography (pQCT) have led to a better assessment of bone development and, in the future, will provide a better understanding of physical activity and inactivity and nutritional status on bone health (21). Variation in body composition associated with sex, ethnicity, and age is well described by Malina (26).

Several methods can be used to estimate muscle mass in children and adults (27). The need for more practical, convenient, and valid methods of estimating muscle mass is evident from the large number of laboratory methods and small number of field methods.

Accurately measuring changes in muscle mass and bone development during growth in children is an important area to be studied. The effect of exercise on bone development has yet to be clearly elucidated. Skeletal changes through the life span can be more completely studied with this new pQCT technology (28). For assessing fat, BMI is widely used but has been shown to be less accurate for estimating body fat in the individual. Skinfolds, ultrasound, and BIA are more accurate methods to be used in body composition field studies, such as Pathways (21) and Trial of Activity for Adolescent Girls (TAAG) (29).

The association of high birth weight and rapid weight gain with childhood and adult obesity is an important area under investigation in recent years. Several studies have shown a relationship between 2- and 3-year-olds and obesity for those who grow more rapidly in the first 6 months and 1 year (10,11,30). Most of the studies use weight for length as a measure of obesity in the first 3 years. Other methods for assessing body composition changes associated with growth in infants include skinfolds (31) and ADP (32). More research is needed on studying the impact of early growth on body composition. The study of tracking body fat in children into adulthood has been investigated by Guo and Chumlea (33). Three-year-old obese children in general do not track well into adulthood. Thus, the long-term effect of infant rapid weight gain on adult body composition is not well established and needs further investigation.

Other Factors Influencing Body Composition

Nutritional status and body composition are also linked through the life cycle as dietary intake changes, and there are expected differences between genders and in health versus disease. A typical nutritional issue in older adults is the fact that diets are low in protein. In fact, current research may lead to changes in the Recommended Dietary Allowance (RDA) for protein in older adults (ages 65 years and older) to prevent sarcopenia (34). There are other physical changes with aging secondary to changes in muscle. For example, as muscle mass decreases, body composition changes in favor of body fat. These changes in body composition can increase the risk of chronic disease, including diabetes and heart disease, but may also lead to bone loss and osteoporosis (34). Thus, assessing body composition in older adults can be helpful because it provides a tool to diagnose both sarcopenia and body fatness, which allows for more targeted interventions relative to disease prevention or treatment. In older adults, body composition is best assessed using DXA because it allows for bone mineral density testing at the lumbar spine and hip to evaluate the risk of osteoporosis. Thus, body composition testing, especially looking for lean components in older adults, may provide information on energy balance and protein intake, whereas bone mass can indicate issues with calcium balance and vitamin D status.

Chronic diseases that cause body wasting (cachexia) can also link body composition to nutritional status. Cachexia is characterized by loss of muscle with or without loss of fat mass. Cachexia is associated with chronic illnesses such as heart failure, chronic obstructive pulmonary disease, kidney disease, infection and sepsis, cancer, and human immunodeficiency virus (35). Whether further body composition testing is warranted in these cases and what method is used is a physician's choice. Simple ways to maintain some information about muscle mass may be through the use of arm muscle area. In hospitalized patients, where chronic periods of both insufficient calorie or protein intake are common (cachexia), arm muscle area is a widely accepted and clinically practical anthropometric method to assess nutritional status. Arm muscle area is calculated using triceps skinfold thickness and midarm circumference (36).

Finally, in undernutrition, 85% of the variance in body weight can be explained by changes in body composition (37). This opens the opportunity to use body composition assessment to determine the degree of undernutrition. But how lean is too lean? Over the years, minimum body fat levels, corresponding to fitness and sport, have been cited as 5% and 12%, in men and women, respectively (2). One of the issues with setting such levels is that each method is characterized by variability, especially if not standardized (38), and athletes exhibit various levels of genetic disposition to maintain health despite excessive thinness. Besides body fat levels, BMI has also been used to scrutinize undernutrition. A level of 18.5 kg/m² has been repeatedly used as a threshold for underweight in clinical and undernourished (39) and athletic populations (40,41).

Accurate body composition assessment in extreme leanness is known to be problematic. Using newer technologies such as ultrasound, which measures fat directly underneath the skin, may prove useful in the future (42,43). Further, refeeding after undernutrition poses a threat to body fat distribution, and using body composition assessments during this phase might be of great value. We touch on this more later in the chapter, with particular interest in the recovery from eating disorders.

Competitive Sports and Exercise Training

All sports and positions played within the same sport have unique desired body types and compositions. Some athletes may self-select toward certain sports based on their inherited genotypes and somatotypes. Some competitive athletes may try achieving their desired body composition for their sports, positions, or weight categories via manipulating training routines, dietary intakes, and other means. Some methods are too extreme, and they often result in negative consequences (40,41). For example, fasting or extended starvation results in energy and nutrient deficiency, with the consequence of poor performance from glycogen depletion. Dehydration with low water intake and use of sweat suits can lower performance and cause metabolic abnormalities. More extreme weight loss methods can even lead to dangerously low levels of electrolytes, especially after self-induced vomiting, which results in large fluid and electrolyte shifts (41).

Caloric restriction is a common method to cut weight. Rapid weight loss with extreme caloric restriction can be associated with greater loss of lean body mass. Garthe and colleagues (44) have shown that gradual weight loss can accomplish minimum mean body mass reduction while reducing weight and fat mass effectively. Chronic dieting and weight cycling are known to reduce metabolic rate, which can affect body composition and make it more difficult to maintain a normal body mass and composition in the long term (45).

Sport dietitians or practitioners typically recommend higher protein intakes while in the weight management phase. The recommendation for weight-sensitive sports is 1.4 to 2.0 g · kg⁻¹ · day⁻¹ (45), whereas higher protein ingestion of 2.3 g · kg⁻¹ · day⁻¹ is noted to preserve lean body mass while on a hypoenergetic diet (46). Further, a recent article by Helms et al. (47) notes protein needs for lean, resistance-trained individuals who are on caloric restriction are 2.3 to 3.1 g · kg fat-free (FFM) mass^{-1} · day^{-1} to help reduce skeletal muscle loss.

Changes in body composition reflect a positive physiological adaptation of exercise training, including alteration in fat mass, lean mass, and bone mineral density. The rate at which such adaptations occur depends on the modality, frequency, intensity, and duration of the exercise (48). Resistance training is known to stimulate mixed muscle protein synthesis, and the anabolic pathway is optimized when combined with proper nutrition and recovery (49,50). The body's ability to adapt to exercise resistance results in muscle hypertrophy, especially in force-generating type II muscle fibers (50). High-intensity training can induce hypertrophic responses similar to those for type IIA muscle fibers, as seen in resistance exercise (1). Weight-bearing exercise and high-impact activity can increase bone mineral density (51,52), and bone mineral density and lean body mass are expected to be higher with resistance-trained populations compared to sedentary groups. Endurance exercise promotes capillarization, mitochondrial density in muscle, and glycogen and fat-storing capability in muscle to increase oxidative capacity rather than substantial hypertrophy (1). Individual responses to exercise depend on many factors, including genotypes. Some genes are identified to intervene with training adaptation, including fat mass, lean mass, bone mineral content, and changes in body composition (53).

When considering the frequency of body composition assessment, Meyer and colleagues (38) recommend testing body composition using DXA no more than two times per year, whereas for skinfolds and other methods, they recommend not to exceed three or four repeated measures in a year. The rationale for capping the frequency of body composition measures using various methods is their limitation to measure changes because a significant weight change is needed to precisely measure a change in body composition (38).

The need to standardize body composition assessment in athletes with regard to methods, tester protocol, measurement frequency, and hydration testing is an important recommendation by the Ad Hoc Working Group of Body Composition, Health and Performance (38) to reduce unhealthy practices related to weight measurement in sport. Further research to standardize body composition methods in sport is recommended (41).

PRACTICAL INSIGHTS

The general use of many methods of body composition to assess body fatness leads to a great deal of variation among laboratories and investigators. Standardization of measurement protocols has led to important work in the area of estimation of minimum weight in many high school and college wrestling programs throughout the country. Both skinfolds and BIA have been widely used in this effort. Future efforts on standardization and training in the field will be important to the increasing application of body composition to the health and performance of athletes.

Body Composition and Eating Disorders

Eating disorders (EDs) are defined in the fifth edition of the *Diagnostic and Statistical Manual of Mental Disorders* (*DSM-5*) as anorexia nervosa (AN), bulimia nervosa (BN), and binge eating disorder (BED) (54). These clinical mental disorders are characterized by abnormal eating behaviors often associated with body image dissatisfaction (55). Because of the wide spectrum of signs and symptoms, the classification and diagnosis of EDs and disordered eating (DE), for example, eating addictions, emotional eating, and restrained eating, is challenging, and distinctive subclinical EDs and DE are also defined (54). EDs and DE often develop at young ages, and, even if females are at higher risk for EDs and DE, male prevalence cannot be neglected (56,57). In a 10-year longitudinal study, Neumark-Sztainer et al. (58) found that the prevalence of dieting and DE was high in adolescent ages and remained constant or increased from adolescence to young adulthood. In the athletic population, a higher prevalence is associated with weight-sensitive sports, such as aesthetic, endurance, and weight-class sports (59).

The consequences of an ED are broad and differ based on the idiosyncrasies of each eating disorder and individual characteristics. Generally, they affect a wide-ranging number of systems, such as the cardiovascular, gastrointestinal, and hormonal systems (60). As a consequence, a holistic approach should be considered during the recovery from an ED.

Body mass and related BMI as well as body composition are common parameters used to support the nutrition care process of assessment, diagnosis, intervention, monitoring, and reassessment, the latter being critical during stages of recovery from an ED.

Patients with differing ED pathology will show diverse body weight ranges. AN is associated with low body mass, BMI, and percentage of body fat, whereas BED is related to overweight or obesity. In cases of extreme low body mass, absolute weight reduction to less than 55% to 60% of ideal body weight might represent a life-threatening risk for the individual. In such cases, refeeding is aggressive, most likely exceeding weekly body mass gain of ~1 to 1.4 kg (~2-3 lb), as commonly recommended in inpatient ED treatment settings (61). It is important to note that fast refeeding strategies, although leading to satisfactory achievement of minimum weight and BMI thresholds and early discharge, may cause more frequent relapses (62,63). Lund et al. (62) therefore recommend a minimum discharge BMI of 20 kg/m^2 to optimize outcomes of treatment and recovery. Moreover, the number of calories needed for weight gain will likely change over treatment and recovery phases as body mass increases (64).

Monitoring body mass and BMI during recovery from an ED, however, might not be enough because neither measurement describes changes in body composition. An increase in percent body fat is of obvious benefit and recommended during ED treatment, especially considering the role of fat in normalizing reproductive function (64). In anorectic patients, 50% of body mass gained is generally in fat mass (65). One of the major issues related to body dissatisfaction and postrecovery relapse in AN is that rapid body mass gains increase fat mass, especially in the abdominal and triceps area (66). To ensure slower body mass gains and more even body fat distribution during recovery, the inclusion of physical exercise, especially resistance exercise training, is recommended to promote lean body mass gain (67,68).

Quite contrary to such recommendations, physical exercise is often prohibited in ED treatment settings, especially in AN and BN patients, to avoid compensatory mechanisms to increased food intake (e.g., the ED patient may tend to increase physical exercise to burn more calories, which would make it harder to meet positive energy balance goals and weight increase). However, including physical activity, and especially strength training, exerts a positive effect on recovery from ED, and it may strengthen both muscle and bone mass (69). Bone mineral density is commonly low in ED patients (70) and may not be reversible in very severe cases with AN. In addition, inpatient cases are also at risk for low vitamin D status (71), underlying issues of insufficient bone mineralization. Thus, ensuring sufficient calcium intake and a normalized vitamin D status; promoting body mass, muscle, and fat gains according to a well-paced recovery strategy; and monitoring the resumption of reproductive function are important aspects of a comprehensive approach that may help ED patients in their journey back to a normal life, where body and mind can live positively in concert.

Fast body mass and fat gains may increase an already distorted view of thinness and will present a major obstacle to recovery without relapse in ED patients. Monitoring body composition is thus important in patients recovering from EDs to minimize disproportionate abdominal fat gains and optimize short- and long-term treatment strategies. As discussed previously, physical exercise, and especially resistance exercise, could favorably influence body composition while in positive energy balance and instill body satisfaction with positive progression and limited relapse risk in patients recovering from an ED.

Mental disorders related to body image acceptance prompt precautions in practitioners to minimize psychological stress when considering body composition assessment. DXA may offer the best approach because it also assesses bone mass as well as lean and fat mass (42).

Body Composition and Weight Loss

With a majority of the population weighing in as "overweight" or "obese," there has been an increased focus on weight loss. Weight loss has been associated with many beneficial health outcomes, such as lowered body fat, which results in improved insulin sensitivity, blood lipids, and glucose levels and lowered risk for a variety of diseases. However, weight loss may increase the risk for mortality and other adverse cardiometabolic outcomes, particularly in the context of weight cycling. Weight loss has also been found to induce adverse effects on lean body mass and bone density (72). The most accurate measurement technique to assess fat mass and FFM is the four-component model because it takes into account water and mineral content. In people with high amounts of body fat, FFM has a higher water content compared to people with lower amounts of fat (under 30%), which may be confounded during weight loss. In older age, myriads of metabolic changes are occurring: fat mass increases (particularly in the abdominal region), and FFM and bone density decrease. Although DXA tends to correlate best with the four-component model (compared to total body water and densitometry), none of the three methods (DXA, total body water, and densitometry), when used by itself (two-component model), measures body composition changes accurately.

Most clinicians and interventionists employ the use of techniques that generally assess fat mass and FFM. However, visceral fat assessment is a critical fat depot that needs to be assessed because it significantly relates to disease. It is also difficult to assess accurately (73). It is well known that magnetic resonance imaging (MRI) and computed tomography (CT) are direct imaging techniques that are the gold standards for measuring visceral fat. However, these tools are mostly inaccessible. Both techniques require expensive equipment and highly trained specialists to read and interpret the results, which may not always be available in most places. It has been recommended that retrospective analyses of visceral fat areas may be desirable within routine clinical care where these methods are employed in evaluation of other diseases (73). MRI is preferable to CT because it does not expose the individual being measured to ionizing radiation (42).

DXA, ADP, BIA, and ultrasound are tools that have been used to assess visceral fat and total fat. These tools are better able to assess total body fat but are not as accurate for measuring visceral fat compared to the CT or MRI. DXA and ADP require the use of formulas to assess visceral fat, though at least one study found very good correlation between CT and DXA for measuring visceral fat in people of a variety of BMIs (74).

Ultrasound may be able to assess visceral fat and total fat but is largely dependent on measurer technique and skill. Some studies have ascertained that it is a relatively good measurement tool for visceral fat in lieu of direct imaging techniques, with correlations ranging between 0.8 and 0.9 (75,76). It is also considered a valid and reliable tool for measuring body fat (77). It is not as well known as other tools, but its high portability and relative inexpensiveness make it an attractive option.

BIA is more affordable than the other tools and more widely available. However, it is also prone to error from hydration status and time of day of measurement. Equations are also applied with BIA to estimate visceral fat (78). Accuracy can be improved if a BIA measurement is obtained in the late afternoon and early evening with the patient being adequately hydrated if total body fat is being measured. BIA is considered one of the less accurate methods compared to ultrasound, waist circumference, and other more direct methods to assess visceral fat (73).

Anthropometrics and field techniques are of much interest in estimating body composition changes because they are inexpensive and easy to use. However, they are prone to large measurement errors without expert training. BMI, for example, is not considered an acceptable tool to assess healthy weight changes in individuals, particularly in active people (79). BMI takes into account only height and weight.

Percent fat, an improved measure compared to BMI, can be obtained using skinfolds. Skinfold measurements are employed by assessing the thickness of skinfolds at certain body sites and the use of equations to estimate percent body fat. They will assess only subcutaneous fat and are prone to error because they are dependent on the skill of the measurer, the number of body sites used for measurement, and the equation used to assess percent body fat. The American College of Sports Medicine has developed a set of specific guidelines for using skinfolds that reduce inter- and intrameasurer error (80).

Body fat distribution is the most desirable measure when trying to assess disease risk because it attempts to estimate visceral fat (81). Although no anthropometric technique can accurately measure visceral fat, it can still provide a good assessment of disease risk. Several popular techniques used include waist circumference, waist-to-height ratio, and waist-to-hip ratio, each with its strengths and weaknesses, but all of them are equally predictive of cardiovascular and diabetes risk (82). Waist-to-hip ratio provides a good snapshot of people's relative body fat distribution and can classify them as "apples" or "pears." Apples are individuals with a preponderance of abdominal (android) fat; they are at higher disease risk than pears, who have a preponderance of gynoid, or subcutaneous, fat predominantly located in the hip region. More recently, the use of multiple anthropometrics is being explored in providing a more accurate assessment of disease risk (83,84). Waist circumference and waist-to-height ratio perform better at capturing weight loss changes than does the waist-to-hip ratio (82). However, the location of measuring waist circumference needs to be standardized because there are several ways it can be measured; yet similar cut points have been suggested for various sites of measurement.

We recommend measuring waist circumference (chapter 4) at the narrowest circumference and midway between the iliac crest and lowest rib. It has been recommended that the waist circumference should be assessed as an integral measurement during weight loss in individuals (85).

PRACTICAL INSIGHTS

If waist circumference is used in the weight loss program along with BIA or skinfolds, it is important to measure regularly marking the 5%, 10%, and 15% weight loss goals. Circumferences and skinfold thicknesses are useful to track changes over time at a given site before whole-body percent fat changes are evident. BIA or skinfolds are useful to determine whole-body percent fat and the changes experienced over time as a person undertakes a weight loss program. Skill in measuring the skinfolds and circumferences is needed, and it is important for the health care provider or trainer to establish the intraobserver error for each site. Large intraobserver errors make it difficult to assess body composition changes with weight loss.

At present, there are few studies documenting the composition of weight loss and weight regain using the four-component model. One excellent study found the effect of using densitometry, DXA, or body water alone using the two-component model leads to over- or underestimation of the body composition changes (86). The investigators found changes in fat mass were underestimated by body water changes and overestimated by densitometry. The accuracy of the DXA estimates was dependent on initial fatness (86).

With weight loss, the amount of lean mass lost is variable, and research has suggested that a variety of factors contribute to this phenomenon: gender, age, sleep, stress, baseline fat mass, macronutrient composition of the diet, energy intake, metabolic state, hormonal responsiveness, and exercise (87). Overall, a more favorable body composition (more fat loss and better preservation of lean mass) can be attained with weight-loss protocols when protein is consumed at higher than recommended levels and resistance training is part of the weight-loss program (46,88-90).

Backx et al. (91) found that increased protein intake (1.7 g/kg) in a group of overweight elderly adults who underwent a 12-week weight-loss program showed no improvement in preservation of lean mass, strength, or physical performance. They also found the losses were less in the high-protein group compared to the normal-protein group (0.9 g/kg) (91). Concerning diet overall and fat loss, studies show that a diet with a low glycemic index (or lower carbohydrate) and higher in protein (compared to fat) has a more favorable effect on fat mass and intraabdominal fat, satiety, thermogenesis, and preservation of lean mass during weight loss than does a diet with a higher glycemic index and higher in carbohydrate (92,93).

It is well known that bone loss and increased bone turnover occur during weight loss (94-96). Although it is not well understood, theories have been proposed that bone turnover and loss during weight loss is a complex interaction between immune function, inflammatory status, and endocrine function (97). In addition, adequate dairy intake may be important in preservation of some bone density during weight loss (95,97). Dairy is not only rich in protein and calcium but may also preserve appetite.

In terms of measurement, weight-loss programs should incorporate both adiposity and abdominal fat assessments. There are many tools for assessing changes, though gold standard tools are limited by inaccessibility. Field tools, such as anthropometrics, may not be as accurate. Using more than one anthropometric tool in conjunction with others may improve the overall metabolic picture as an individual loses weight. The use of body water, DXA, or densitometry (underwater weighing or ADP) and the two-component model leads to biased estimates of fat and FFM changes. At present only the four-component model and MRI can be used to obtain accurate measures of body composition changes.

Body Composition, Chronic Disease, and Aging

Body composition plays a determinant role in health status in the last decades of life. Body composition changes occur with the aging process, and much of what we know comes from cross-sectional and longitudinal data, with some caution in the interpretation of this information. When using cross-sectional data, body composition changes are estimated from one cohort of persons who differ in age; however, extrapolation warrants a careful interpretation because there can be an effect of the birth cohort on body composition (98). That is, persons born in the 1930s may differ initially in their body composition from those born in the 1970s. The birth cohort effect should also be considered when using data sets of older individuals as a reference because there are secular changes that may confound the comparison. On the other hand, currently available longitudinal data arise from selected samples that are not population based and therefore may not be valid in other contexts.

Longitudinal data illustrate that older adults' percent fat mass increases with age, leveling off at about 80 years of age, whereas absolute fat mass increases until 80 years of age and then decreases, with this reduction being more rapid in males compared to females (98). Excess adiposity in older adults is related to an increased likelihood of functional limitations and is also a major risk factor for a proinflammatory profile (99,100). Changes in body fat distribution take place with the aging process and are determinants for health status (101). Franklin et al. (102) observed that both subcutaneous and visceral fat increase through menopause. Longitudinal data show relative visceral adipose tissue increases of 0.22% per year in males and 0.27% per year in females (103). When considering waist circumference, age-associated increases are also greater in females (0.28 cm [0.11 in.] per year) than in males (0.18 cm [0.07 in.] per year) (104).

Appendicular lean mass declines by about 0.4 kg (0.9 lb) per decade in females and 0.8 kg (1.8 lb) per decade in males (105). Senescence is associated with reductions in muscle cross-sectional area and changes in the number and type of muscle fibers (106,107). The rate of age-related muscle mass decrease is accelerated from 60 years of age, which may impair muscle function and muscle power (107). The loss of skeletal muscle mass is a major concern in the elderly and has been associated with disability, comorbidities, and a higher risk of mortality in older people.

More than the age-associated decline in lean mass or obesity alone, a combination of obesity and muscle impairment (defined as low muscle mass or poor muscle strength), designated as sarcopenic obesity, has been established as a focus of attention in the elderly. In older adults, an association between cardiometabolic variables and sarcopenic obesity has been identified (108). There is a possible role of obesity-associated inflammation in the age-related process that pertains to the development and progression of sarcopenic obesity (109). More than sarcopenia or obesity alone, lower muscle mass and strength and greater fat infiltration into the muscle have been associated with a disability and incident mobility limitations (110,111).

Research with risk factors for cancer include both body fatness and visceral fat. Overall fatness increases inflammation, and visceral fat adds additional risk through its association with insulin resistance (112). The protective effect of certain fat depots on decreasing chronic disease adds a complexity to the relation of body fat to health and disease.

In older ages, changes that occur in bone mineral density are also a focus of attention. Cross-sectional information shows that bone mineral density declines with age in both cortical and trabecular bones and that this loss is greater in females (113). Over a 10-year follow-up period, there are differences in femoral neck bone mineral density changes among normal-weight, overweight, and obese older adults but no BMI category differences in total hip and whole-body bone mineral density changes over time (114). Low bone mineral density has been associated with the incidence of cardiovascular diseases (115). According to the World Health Organization (116), osteoporosis is defined by a bone mineral density (at the femoral neck, total hip, or lumbar spine) ≤2.5 standard deviations below the average value for young, healthy women (T-score ≤2.5 standard deviations). Age-related bone loss is asymptomatic; still, osteoporosis causes more than 8.9 million fractures each year worldwide and causes people to become incapacitated, disabled, and physically dependent (117).

Considering the health-related changes that occur in body composition with aging, it is fundamental to accurately estimate older adults' body components. Overall, body composition methods that are used in adults may be applicable to older adults; still, there are some considerations that should be taken into account. Basic two-component molecular models are based on the assumption that FFM density is 1.1 g/cm^3 (0.04 lb/in.^{3}) (densitometric models) (118) or that FFM hydration is 73.2% (hydrometric models) (119). There has been some controversy about whether older adults may deviate from this assumed constant and there may be individual differences in the FFM density and composition (120-123). The lack of consensus may be related to different sample characteristics in older ages. There is an increased prevalence of numerous diseases and medications with aging that may affect body composition; accordingly, the disparity between studies may reflect these different characteristics rather than the effect of age. Nevertheless, caution is necessary when using two-component models in older adults. By including more and different measured properties or other components, three- and four-component models typically account for more biological variability (124).

Considering body composition assessment with DXA, there may be small systematic but predictable errors in body composition estimations from soft tissue hydration variations (125). Still, DXA is considered a precise, convenient, and useful diagnostic tool for body composition estimations. Additionally, DXA provides the possibility of regional analysis, which allows a better comprehension of body fat distribution (126). Furthermore, in older adults, DXA is of greater value for the early diagnosis of osteopenia and osteoporosis (116).

An excellent example of the use of DXA is in the Health ABC (Health, Aging and Body Composition) Study, where over 3,000 individuals were followed for 3 years to better describe the changes in function and composition in older Americans. One of the findings by Viser et al. (127) indicates that fat mass is a better predictor of lower extremity performance in older women than in men and that interventions to decrease physical disability through changes in body composition may have different emphasis in older women versus men. Additional studies by Goodpaster et al. (128) and Newman et al. (129) found weight loss further affects strength decreases in aging adults. Another example of DXA use is in the weight-training study of Cussler et al. (130) with postmenopausal women and in the study by Milliken et al. (131), where 4-year changes in both lean mass and bone density are associated with increased weights lifted.

Apart from mechanistic methods, several mathematical models (type I) are available for body composition estimates in older adults. When using type I methods, it is mandatory to choose methods that are population specific to older adults. Accordingly, if a method is to be applied in older adults, it has to be developed and cross-validated in the elderly (132).

Body fat can be estimated from anthropometric equations that include circumferences (133) or skinfolds (134). Anthropometric equations that are used in older adults have in common the fact that age is considered a significant predictor and a positive coefficient is attributed to this variable. This means that, for the same anthropometric

PRACTICAL INSIGHTS

Weight loss in older adults can lead to significant decreases in both muscle mass and bone density. Therefore, it is important to measure changes in these two components using DXA as a measure of both bone density (hip and spine) and lean soft tissue of the arms, trunk, and legs. While the former is performed routinely as a screening for osteoporosis, the latter is not typically performed. This test is easy to perform and could be done at the same visit as the osteoporosis screening scan. Health care professionals would need some training in how to interpret the results, but this would give valuable information about whole-body lean mass changes that occur under a variety of conditions. With the wide availability of DXA in hospitals and clinics throughout the country, it is not difficult to obtain DXA scans on clients undergoing weight changes or with aging.

values, an older person will have higher levels of adiposity. Also, when using BIA, there are solutions developed for the elderly to estimate body composition (121,135). The use of phase angle (resistance vs. reactance) along with resistance from BIA may provide additional information in older women undergoing resistance training (136).

Other Applications

Other developments in the field of body composition include BIA and its use in medicine, the use of MRI for infant body composition, the use of ultrasound to more objectively estimate body fatness, and investigations into body shape and mortality risk.

One of the applications of BIA is to measure the phase angle, where reactance is plotted against resistance. Different reference norms for phase angle have been published and discussed by Norman and colleagues (137), with a review of applications. A low phase angle may be useful as an indicator of inadequate nutrition status; muscle mass loss; or disease, such as cancer, chronic heart failure, and kidney disease (138). Additional applications of BIA use multifrequency BIA and the ratio of impedance at high (200 kHz) and low (5 kHz) frequencies to diagnose malnutrition (139).

Lukaski et al. (140) present further analysis of assessing adult malnutrition using both phase angle and impedance ratios. The authors summarize research on low phase angle related to hospital patients, cancer, hemodialysis, surgery, critical illness, cirrhosis, hepatitis C virus, chronic obstructive pulmonary disease, neuromuscular disease, and geriatrics. An example of research in the hemodialysis area is the study by Onofriescu et al. (141), which found better control of overhydration with BIA monitoring.

One of the promising new methods to assess body composition in infants and young children is the use of EchoMRI quantitative magnetic resonance (QMR) (142,143). The development of protocols to obtain more accurate estimates of muscle and organ size and regional fat depots has enabled MRI to become a major reference method for infants to better assess body composition during growth.

Ultrasound can also be applied to regional muscle mass estimates to provide clinical monitoring of patient recovery (138). Both BIA and ultrasound technologies can be applied to assessing adult malnutrition at bedside, with additional research needed to evaluate their unique place in medical diagnosis and recovery. New developments in ultrasound software are reviewed in chapter 4, and they offer more objective estimates of subcutaneous fat thickness in the athletic population.

The investigation of body shape and regional body composition using algorithms to process whole-body DXA scans into body thickness and leanness images may predict metabolic health and health outcomes. This area of research is illustrated by the work of Shepherd et al. (144) and offers promise of identifying new phenotypes of different groups that predict mortality risk.

Summary

In this chapter, some of the major applications of body composition to the areas of nutritional status, growth and development, competitive sport, exercise training, eating disorders, weight loss, and aging were covered. Also, application of body composition assessment to infants, malnutrition diagnosis, and bone development in children was reviewed. Use of BIA, ultrasound, and MRI methodologies to enhance body composition assessment in the fields of medicine, sports nutrition, and infant growth studies was described. Assessment of visceral obesity, muscle and bone mineral loss, and metabolic health can enhance nutritional status evaluation. Body composition of the mother before birth and of the growing infants and young children can predict early development of childhood obesity. The effects of dehydration, caloric restriction, and dietary protein levels on physical performance were reviewed. The positive effects of exercise training on various aspects of body composition changes were described. Body composition changes during recovery from eating disorders were reviewed along with the advantages of DXA to quantify successful recovery from eating disorders. Limitations of using the two-component model to assess body composition during weight loss and body composition with aging were described. Multicomponent methods need to be applied to aging and weight-loss populations to better describe the body composition changes taking place.

REFERENCES

Preface

1. Brozek J. *Body Composition.* New York, NY: New York Academy of Sciences; 1963.

2. Roche AF, Heymsfield SB, Lohman TG. *Human Body Composition.* Champaign, IL: Human Kinetics; 1996.

3. Heymsfield SB, Lohman TG, Wang Z, Going SB. *Human Body Composition.* 2nd ed. Champaign, IL: Human Kinetics; 2005.

4. Heyward V, Stolarczyk L. *Applied Body Composition Assessment.* Champaign, IL: Human Kinetics; 1996.

Chapter 1

1. Meyer NL, Sundgot-Borgen J, Lohman TG, et al. Body composition for health and performance: a survey of body composition assessment practice carried out by the Ad Hoc Research Working Group on Body Composition, Health and Performance, under the auspices of the IOC Medical Commission. *Br J Sports Med.* 2013;47(16):1044-1053.

2. Ackland T, Lohman T, Sundgot-Borgen J, et al. Current status of body composition assessment in sport: review and position statement on behalf of the Ad Hoc Research Working Group on Body Composition Health and Performance, under the auspices of the I.O.C. Medical Commission. *Sports Med.* 2012;42(3):227-249.

3. Lohman TG, Pollock ML, Slaughter MH, Brandon LJ, Boileau RA. Methodological factors and the prediction of body fat in female athletes. *Med Sci Sports Exerc.* 1984;16(1):92-96.

4. Jackson AS, Pollock ML. Practical assessment of body composition. *Phys Sports Med.* 1985;13(5):76-90.

5. Lohman TG. *Advances in Body Composition Assessment.* Champaign, IL: Human Kinetics; 1992.

6. Bland JM, Altman DG. Statistical methods for assessing agreement between two methods for clinical measurement. *Lancet.* 1986;8:307-310.

7. Behnke AR. Discussion. In: Menecky GR, Linde SM, eds. *Radioactivity in Man.* Springfield, IL: Charles C Thomas; 1965.

8. Brozek J, Grande F, Anderson JT, Keys A. Densitometric analysis of body composition: revision of some quantitative assumptions. *Ann NY Acad Sci.* 1993;110:113-140.

9. Lohman TG. Assessment of body composition in children. *Pediatr Exerc Sci.* 1989;1:19-30.

10. Ellis, KJ. Reference man and woman more fully characterized. In: Zeisler R, Guinn VP, eds. *Nuclear Analytical Methods in the Life Sciences.* Totowa, NJ: Human Press; 1990.

Chapter 2

1. Wang Z-M, Pierson RN Jr, Heymsfield SB. The five-level model: a new approach to organizing body-composition research. *Am J Clin Nutr.* 1992;56:19-28.

2. Selinger A. *The Body as a Three Component System* [dissertation]. Champaign: University of Illinois Urbana–Champaign; 1977.

3. Lohman TG. Applicability of body composition techniques and constants for children and youth. *Exerc and Sport Sci Rev.* 1986;14:325-357.

4. Snyder WS, Cook MJ, Nasset ES, Karhausen LR, Howells GP, Tipton IH. *Report on the Task Group on Reference Man.* Oxford, UK: Pergamon Press; 1984.

5. Heymsfield SB, Waki M, Kehayias J, et al. Chemical and elemental analysis of humans in vivo using improved body composition models. *Am J Physiol.* 1991;261:E190-E198.

6. Behnke AR, Wilmore JH. *Evaluation and Regulation of Body Build and Composition.* Englewood Cliffs, NJ: Prentice Hall; 1974.

7. Forbes GB. *Human Body Composition: Growth, Aging, Nutrition, and Activity.* New York, NY: Springer-Verlag; 1987.

8. Moore FD, Olesen KH, McMurray JD, Parker HV, Ball MR, Boyden CM. *The Body Cell Mass and Its Supporting Environment.* Philadelphia, PA: Saunders; 1963.

9. Lohman TG, Roche AF, Martorell R. *Anthropometric Standardization Reference Manual.* Champaign, IL: Human Kinetics; 1988.

10. Siri WE. Body composition from fluid spaces and density: analysis of methods. In: Brozek J, Henschel A, eds. *Techniques for Measuring Body Composition.* Washington, DC: National Academy of Sciences; 1961:223-244.

11. Heyward VH. Practical body composition assessment for children, adults, and older adults. *Int J Sport Nutr.* 1998;8(3):285-307.

12. Brozek J, Grande F, Anderson JT, Keys A. Densitometric analysis of body composition: revision of some quantitative assumptions. *Ann NY Acad Sci.* 1963;110:113-140.

13. Visser M, Gallagher D, Deurenberg P, Wang J, Pierson RN Jr., Heymsfield SB. Density of fat-free body mass: relationship with race, age, and level of body fatness. *Am J Physiol.* 1997;272:E781-E787.

14. Roemmich JN, Clark PA, Weltman A, Rogol AD. Alterations in growth and body composition during puberty: comparing multicompartment body composition models. *J Appl Physiol.* 1997;83(3):927-935.

15. Streat SJ, Beddoe AH, Hill GL. Measurement of body fat and hydration of the fat-free body in health and disease. *Metabolism.* 1985;34(6):509-518.

16. Modlesky CM, Cureton KJ, Lewis RD, Prior BM, Sloniger MA, Rowe DA. Density of the fat-free mass and estimates of body composition in male weight trainers. *J Appl Physiol.* 1996;80(6):2085-2096.

17. Prior BM, Modlesky CM, Evans EM, et al. Muscularity and the density of the fat-free mass in athletes. *J Appl Physiol.* 2001;90(4):1523-1531.

18. Wang Z, Shen W, Withers RT, Heymsfield SB. Multicomponent molecular-level models of body composition analysis. In: Heymsfield SB, Lohman TG, Wang ZM, Going SB, eds. *Human Body Composition.* 2nd ed. Champaign, IL: Human Kinetics; 2005:163-175.

19. Wang Z. High ratio of resting energy expenditure to body mass in childhood and adolescence: a mechanistic model. *Am J Hum Biol.* 2012;24(4):460-467.

20. Wang Z, Zhang J, Ying Z, Heymsfield SB. New insights into scaling of fat-free mass to height across children and adults. *Am J Hum Biol.* 2012;24(5):648-653.

21. Wang Z, Deurenberg P, Guo S, et al. Six-compartment body composition model: inter-method comparisons of total body fat measurement. *Int J Obes.* 1998;22(4):329-337.

22. Ellis KJ. Whole-body counting and neutron activation analysis. In: Heymsfield SB, Lohman TG, Wang ZM, Going SB, eds. *Human Body Composition.* 2nd ed. Champaign, IL: Human Kinetics; 2005:51-62.

23. Ross R, Janssen I. Computed tomography and magnetic resonance imaging. In: Heymsfield SB, Lohman TG, Wang ZM, Going SB, eds. *Human Body Composition.* 2nd ed. Champaign, IL: Human Kinetics; 2005:89-108.

24. Ellis KJ. Human body composition: in vivo methods. *Physiol Rev.* 2000;80(2):649-680.

25. Heymsfield S, Lohman TG, Wang ZM, Going SB, eds. *Human Body Composition.* 2nd ed. Champaign, IL: Human Kinetics; 2005.

26. Engstrom CM, Loeb GE, Reid JG, Forrest WJ, Avruch L. Morphometry of the human thigh muscles: a comparison between anatomical sections and computer tomographic and magnetic resonance images. *J Anat.* 1991;176:139-156.

27. Abate N, Burns D, Peshock RM. Estimation of adipose tissue mass by magnetic resonance imaging: validation against dissection in human cadavers. *J Lipid Res.* 1994;35:1490-1496.

28. Mitsiopoulos N, Baumgartner RN, Heymsfield SB, Lyons W, Gallagher D, Ross R. Cadaver validation of skeletal muscle measurement by magnetic resonance imaging and computerized tomography. *J Appl Physiol.* 1998;85(1):115-122.

29. Hu HH, Li Y, Nagy TR, Goran MI, Nayak KS. Quantification of absolute fat mass by magnetic resonance imaging: a validation study against

chemical analysis. *Int J Body Compos Res.* 2011;9(3):111-122.

30. Shen W, Wang Z, Tang H, et al. Volume estimates by imaging methods: model comparisons with visible woman as the reference. *Obes Res.* 2003;11(2):217-225.

31. Ross R, Goodpaster B, Kelley D, Boada F. Magnetic resonance imaging in human body composition research: from quantitative to qualitative tissue measurement. *Ann N Y Acad Sci.* 2000;904:12-17.

32. Gallagher D, Heymsfield SB. Muscle distribution: variations with body weight, gender, and age. *Appl Radiat Isot.* 1998;49(5-6):733-734.

33. Farr JN, Funk JL, Chen Z, et al. Skeletal muscle fat content is inversely associated with bone strength in young girls. *J Bone Miner Res.* 2011;26(9):2217-2225.

34. Farr JN, Van Loan MD, Lohman TG, Going SB. Lower physical activity is associated with skeletal muscle fat content in girls. *Med Sci Sports Exerc.* 2012;44(7):1375-1381.

35. Goodpaster BH, Thaete FL, Kelley DE. Thigh adipose tissue distribution is associated with insulin resistance in obesity and in type 2 diabetes mellitus. *Am J Clin Nutr.* 2000;71(4):885-892.

36. Goodpaster BH, Thaete FL, Simoneau JA, Kelley DE. Subcutaneous abdominal fat and thigh muscle composition predict insulin sensitivity independently of visceral fat. *Diabetes.* 1997;46:1579-1585.

37. Chen Z, Wang Z, Lohman T, et al. Dual-energy X-ray absorptiometry is a valid tool for assessing skeletal muscle mass in older women. *J Nutr.* 2007;137(12):2775-2780.

38. Wang Z-M, Visser M, Ma R, et al. Skeletal muscle mass: evaluation of neutron activation and dual-energy X-ray absorptiometry methods. *J Appl Physiol.* 1996;80(3):824-831.

Chapter 3

1. Brozek J, Grande F, Anderson JT, Keys A. Densitometric analysis of body composition: revision of some quantitative assumptions. *Ann N Y Acad Sci.* 1963;110:113-140.

2. Keys A, Brozek J. Body fat in adult man. *Physiol Rev.* 1953;33:245-325.

3. Siri WE. Body composition from fluid spaces and density: analysis of methods. In: Brozek J, Henschel A, eds. *Techniques for Measuring Body Composition.* Washington, DC: National Academy of Sciences; 1961:223-244.

4. Brodie D, Moscrip V, Hutcheon R. Body composition measurement: a review of hydrodensitometry, anthropometry, and impedance methods. *Nutrition.* 1998;14(3):296-310.

5. Lohman TG. Skinfolds and body density and their relation to body fatness: a review. *Hum Biol.* 1981;53(2):181-225.

6. Hansen NJ, Lohman TG, Going SB, et al. Prediction of body composition in premenopausal females from dual-energy X-ray absorptiometry. *J Appl Physiol.* 1993;75(4):1637-1641.

7. Kohrt WM. Preliminary evidence that DEXA provides an accurate assessment of body composition. *J Appl Physiol.* 1998;84(1):372-377.

8. Going SB. Hydrodensitometry and air displacement plethysmography. In: Heymsfield SB, Lohman TG, Wang ZM, Going SB, eds. *Human Body Composition.* 2nd ed. Champaign, IL: Human Kinetics; 2005:17-33.

9. Ackland T, Lohman T, Sundgot-Borgen J, et al. Current status of body composition assessment in sport: review and position statement on behalf of the Ad Hoc Research Working Group on Body Composition Health and Performance, under the auspices of the I.O.C. Medical Commission. *Sports Med.* 2012;42(3):227-249.

10. Buskirk ER. Underwater weighing and body density: a review of procedures. In: Brozek J, Henschel A, eds. *Techniques for Measuring Body Composition.* Washington, DC: National Academy of Sciences, National Research Council; 1961:90-105.

11. Wilmore JH. A simplified method for determination of residual lung volumes. *J Appl Physiol.* 1969;27(1):96-100.

12. Akers R, Buskirk ER. An underwater weighing system utilizing "force cube" transducers. *J Appl Physiol.* 1969;26(5):649-652.

13. Heymsfield S, Lohman TG, Wang ZM, Going SB, eds. *Human Body Composition*. 2nd ed. Champaign, IL: Human Kinetics; 2005.

14. Jackson AS, Pollock ML, Graves JE, Mahar MT. Reliability and validity of bioelectrical impedance in determining body composition. *J Appl Physiol*. 1988;64(2):529-534.

15. Durnin JV, Satwanti. Variations in the assessment of the fat content of the human body due to experimental technique in measuring body density. *Ann Hum Biol*. 1982;9(3):221-225.

16. Girandola RN, Wiswell RA, Romero G. Body composition changes resulting from fluid ingestion and dehydration. *Res Q Exerc Sport*. 1977;48(2):299-303.

17. Brodie DA, Eston RG, Coxon AY, Kreitzman SN, Stockdale HR, Howard AN. Effect of changes of water and electrolytes on the validity of conventional methods of measuring fat-free mass. *Ann Nutr Metab*. 1991;35(2):89-97.

18. Sinning WE. Body composition assessment of college wrestlers. *Med Sci Sports*. 1974;6(2):139-145.

19. Gnaedinger RH, Reineke EP, Pearson AM, Vanhuss WD, Wessel JA, Montoye HJ. Determination of body density by air displacement, helium dilution, and underwater weighing. *Ann N Y Acad Sci*. 1963;110:96-108.

20. Taylor A, Aksoy Y, Scopes JW, du Mont G, Taylor BA. Development of an air displacement method for whole body volume measurement of infants. *J Biomed Eng*. 1985;7(1):9-17.

21. Gundlach BL, Visscher GJW. The plethysmometric measurement of total body volume. *Hum Biol*. 1986;58(5):783-799.

22. Dempster P, Aitkens S. A new air displacement method for the determination of human body composition. *Med Sci Sports Exerc*. 1995;27(12):1692-1697.

23. McCrory MA, Gomez TD, Bernauer EM, Molé PA. Evaluation of a new air displacement plethysmograph for measuring human body composition. *Med Sci Sports Exerc*. 1995;27(12):1686-1691.

24. Demerath EW, Guo SS, Chumlea WC, Towne B, Roche AF, Siervogel RM. Comparison of percent body fat estimates using air displacement plethysmography and hydrodensitometry in adults and children. *Int J Obes Relat Metab Disord*. 2002;26(3):389-397.

25. Fields DA, Goran MI, McCrory MA. Body-composition assessment via air-displacement plethysmography in adults and children: a review. *Am J Clin Nutr*. 2002;75(3):453-467.

26. Sly PD, Lanteri C, Bates JH. Effect of the thermodynamics of an infant plethysmograph on the measurement of thoracic gas volume. *Pediatr Pulmonol*. 1990;8(3):203-208.

27. Ruppell G. *Manual of Pulmonary Function Testing*. St. Louis, MO: Mosby; 1994.

28. Nunez C, Kovera AJ, Pietrobelli A, et al. Body composition in children and adults by air displacement plethysmography. *Eur J Clin Nutr*. 1999;53(5):382-387.

29. Noreen EE, Lemon PW. Reliability of air displacement plethysmography in a large, heterogeneous sample. *Med Sci Sports Exerc*. 2006;38(8):1505-1509.

30. Tucker LA, Lecheminant JD, Bailey BW. Test-retest reliability of the Bod Pod: the effect of multiple assessments. *Percept Mot Skills*. 2014;118(2):563-570.

31. Iwaoka H, Yokoyama T, Nakayama T, et al. Determination of percent body fat by the newly developed sulfur hexafluoride dilution method and air displacement plethysmography. *J Nutri Sci Vitaminol*. 1998;44(4):561-568.

32. Baracos V, Caserotti P, Earthman CP, et al. Advances in the science and application of body composition measurement. *JPEN J Parenter Enteral Nutr*. 2012;36(1):96-107.

33. Snyder WS, Cook MJ, Nasset ES, Karhausen LR, Howells GP, Tipton IH. *Report of the Group on Reference Man*. Oxford, UK: Pergamon Press; 1975.

34. Edelman IS, Olney JM, James AH, Brooks L, Moore FD. Body composition: studies in the human being by the dilution principle. *Science*. 1952;115(2991):447-454.

35. Moore FD. Determination of total body water and solids with isotopes. *Science.* 1946;104(2694):157-160.

36. Keith NM, Rowntree LG, Geraghty JT. A method for the determination of plasma and blood volume. *Arch Internal Med.* 1915;16:547.

37. Pace N, Kline L, Schachman HK, Harfenist M. Studies on body composition; use of radioactive hydrogen for measurement in vivo of total body water. *J Biol Chem.* 1947;168(2):459-469.

38. Schoeller DA, Van Santen E, Peterson WM, Dietz W, Jaspan J, Klein PD. Total body water measurement in humans with ^{18}O and ^{2}H labeled water. *Am J Clin Nutr.* 1980;33:2686-2693.

39. Schoeller DA. Hydrometry. In: Heymsfield SB, Lohman TG, Wang ZM, Going SB, eds. *Human Body Composition.* 2nd ed. Champaign, IL: Human Kinetics; 2005:35-49.

40. Ellis KJ. Human body composition: in vivo methods. *Physiol Rev.* 2000;80(2):649-680.

41. Racette SB, Schoeller DA, Luke AH, Shay K, Hnilicka J, Kushner RF. Relative dilution spaces of 2H- and 18O-labeled water in humans. *Am J Physiol.* 1994;267(4 Pt 1):E585-590.

42. Schoeller DA. Measurement of total body water: isotope dilution techniques. In: Roche AF, ed. *Body Composition Assessment in Youth and Adults: Sixth Ross Conferences on Medical Research.* Columbus, OH: Ross Laboratories; 1985:124-129.

43. Schloerb PR, Friis-Hansen BJ, Edelman IS, Solomon AK, Moore FD. The measurement of total body water in the human subject by deuterium oxide dilution. 1950:1296-1310.

44. Wong WW, Cochran WJ, Klish WJ, Smith EO, Lee LS, Klein PD. In vivo isotope-fractionation factors and the measurement of deuterium- and oxygen-18-dilution spaces from plasma urine, saliva, respiratory water vapor, and carbon dioxide. *Am J Clin Nutr.* 1988;47:1-6.

45. Schoeller DA, Leitch CA, Brown C. Doubly labeled water method: in vivo oxygen and hydrogen isotope fractionation. *Am J Physiol.* 1986;251(6 Pt 2):R1137-1143.

46. Denne SC, Patel D, Kalhan SC. Total body water measurement in normal and diabetic pregnancy: evidence for maternal and amniotic fluid equilibrium. *Biol Neonate.* 1990;57(5):284-291.

47. McCullough AJ, Mullen KD, Kalhan SC. Measurements of total body and extracellular water in cirrhotic patients with and without ascites. *Hepatology.* 1991;14(6):1102-1111.

48. Schoeller DA. Measurement of energy expenditure in free-living humans by using doubly labeled water. *J Nutr.* 1988;118(11):1278-1289.

49. National Research Council. Water and electrolytes. In: *Recommended Dietary Allowances.* 10th ed. Washington, DC: The National Academies Press; 1989.

50. Schoeller DA. Isotope dilution methods. In: Brodoff PB, ed. *Obesity.* New York, NY: Lippincott; 1991:80-88.

51. Forbes GB. Methods for determining composition of the human body. With a note on the effect of diet on body composition. *Pediatrics.* 1962;29:477-494.

52. Moore FD, Olesen KH, McMurray JD, Parker HV, Ball MR, Boyden CM. *The Body Cell Mass and Its Supporting Environment.* Philadelphia: Saunders; 1963.

53. Pace N, Rathburn EN. Studies of body composition. III. The body water and chemically combined nitrogen content in relation to fat content. *J Biol Chem.* 1945;158:685-691.

54. Wang Z, Deurenberg P, Wang W, Pietrobelli A, Baumgartner RN, Heymsfield SB. Hydration of fat-free body mass: new physiological modeling approach. *Am J Physiol.* 1999;276(6 Pt 1):E995-E1003.

55. Wang ZM, Deurenberg P, Wang W, Pietrobelli A, Baumgartner RN, Heymsfield SB. Hydration of fat-free body mass: review and critique of a classic body-composition constant. *Am J Clin Nutr.* 1999;69:833-841.

56. Speakman JR, Nair KS, Goran MI. Revised equations for calculating CO2 production from doubly labeled water in humans. *Am J Physiol.* 1993;264(6 Pt 1):E912-917.

57. Moulton CR. Age and chemical development in mammals. *J Biol Chem.* 1923:79-97.

58. Fomon SJ, Haschke F, Ziegler EE, Nelson SE. Body composition of reference children from birth to age 10 years. *Am J Clin Nutr.* 1982;35(5 Suppl):1169-1175.

59. Lohman TG. Applicability of body composition techniques and constants for children and youths. *Exerc Sport Sci Rev.* 1986;14:325-357.

60. Wells JC, Williams JE, Chomtho S, et al. Pediatric reference data for lean tissue properties: density and hydration from age 5 to 20 y. *Am J Clin Nutr.* 2010;91(3):610-618.

61. Lohman TG. Research progress in validation of laboratory methods of assessing body composition. *Med Sci Sports Exerc.* 1984;16(6):596-605.

62. Martin AD, Drinkwater DT. Variability in the measures of body fat: assumptions or technique? *Sports Med.* 1991;11(5):277-288.

63. Modlesky CM, Cureton KJ, Lewis RD, Prior BM, Sloniger MA, Rowe DA. Density of the fat-free mass and estimates of body composition in male weight trainers. *J Appl Physiol.* 1996;80(6):2085-2096.

64. Prior BM, Modlesky CM, Evans EM, et al. Muscularity and the density of the fat-free mass in athletes. *J Appl Physiol.* 2001;90(4):1523-1531.

65. Silva AM, Fields DA, Quiterio AL, Sardinha LB. Are skinfold-based models accurate and suitable for assessing changes in body composition in highly trained athletes? *J Strength Cond Res.* 2009;23(6):1688-1696.

66. Silva AM, Minderico CS, Teixeira PJ, Pietrobelli A, Sardinha LB. Body fat measurement in adolescent athletes: multicompartment molecular model comparison. *Eur J Clin Nutr.* 2006;60(8):955-964.

67. Withers RT, Noell CJ, Whittingham NO, Chatterton BE, Schultz CG, Keeves JP. Body composition changes in elite male bodybuilders during preparation for competition. *Aust J Sci Med Sport.* 1997;29(1):11-16.

68. Arngrimsson S, Evans EM, Saunders MJ, Ogburn CL 3rd, Lewis RD, Cureton KJ. Validation of body composition estimates in male and female distance runners using estimates from a four-component model. *Am J Hum Biol.* 2000;12(3):301-314.

69. Moon JR, Tobkin SE, Smith AE, et al. Anthropometric estimations of percent body fat in NCAA Division I female athletes: a 4-compartment model validation. *J Strength Cond Res.* 2009;23(4):1068-1076.

70. Penn IW, Wang ZM, Buhl KM, Allison DB, Burastero SE, Heymsfield SB. Body composition and two-compartment model assumptions in male long distance runners. *Med Sci Sports Exerc.* 1994;26(3):392-397.

71. Wang Z-M, Pierson RN Jr, Heymsfield SB. The five-level model: a new approach to organizing body-composition research. *Am J Clin Nutr.* 1992;56:19-28.

72. Heymsfield SB, Wang Z, Baumgartner RN, Ross R. Human body composition: advances in models and methods. *Ann Rev Nutr.* 1997;17:527-558.

73. Withers RT, Laforgia J, Heymsfield SB. Critical appraisal of the estimation of body composition via two-, three-, and four-compartment models. *Am J Hum Biol.* 1999;11(2):175-185.

74. Clasey JL, Kanaley JA, Wideman L, et al. Validity of methods of body composition assessment in young and older men and women. *J Appl Physiol.* 1999;86(5):1728-1738.

75. Lohman TG, Harris M, Teixeira PJ, Weiss L. Assessing body composition and changes in body composition: another look at dual-energy X-ray absorptiometry. *Ann N Y Acad Sci.* 2000;904:45-54.

76. Coward WA. Calculations of pool sizes and flux rates. In: Prentice AM, ed. *The Doubly-Labelled Water Method for Measuring Energy Expenditure: A Consensus Report by the IDECG Working Group.* Vienna: International Dietary Energy Consultancy Group; 1990.

77. Ellis KJ. Whole-body counting and neutron activation analysis. In: Heymsfield SB, Lohman TG, Wang ZM, Going SB, eds. *Human Body Composition.* 2nd ed. Champaign, IL: Human Kinetics; 2005:51-62.

78. Lohman TG. *Advances in Body Composition Assessment.* Champaign, IL: Human Kinetics; 1992.

79. Cohn SH, Parr RM. Nuclear-based techniques for the in vivo study of human body composition: report of an Advisory Group of the International Atomic Energy Agency. *Clin Phys Physiol Meas.* 1985;6(4):275-301.

80. Lohman TG, Norton, HW. Distribution of potassium in steers by 40 >K measurement. *J Animal Science.* 1968;27:1266-1272.

81. Lohman TG, Ball RH, Norton HW. Biological and technical sources of variability in bovine carcass lean tissue composition II. Biological variation in potassium, nitrogen, and water. *J Animal Science.* 1970;30:21-26.

82. Wang Z, Zhu S, Wang J, Pierson RN Jr., Heymsfield SB. Whole-body skeletal muscle mass: development and validation of total-body potassium prediction models. *Am J Clin Nutr.* 2003;77(1):76-82.

83. Lohman TG, Chen Z. Dual-energy X-ray absorptiometry. In: Heymsfield SB, Lohman TG, Wang ZM, Going SB, eds. *Human Body Composition.* 2nd ed. Champaign, IL: Human Kinetics; 2005:63-77.

84. Fuller NJ, Laskey MA, Elia M. Assessment of the composition of major body regions by dual-energy X-ray absorptiometry (DEXA), with special reference to limb muscle mass. *Clin Physiol.* 1992;12(3):253-266.

85. Genton L, Hans D, Kyle UG, Pichard C. Dual-energy X-ray absorptiometry and body composition: differences between devices and comparison with reference methods. *Nutrition.* 2002;18(1):66-70.

86. Toombs RJ, Ducher G, Shepherd JA, De Souza MJ. The impact of recent technological advances on the trueness and precision of DXA to assess body composition. *Obesity (Silver Spring).* 2012;20(1):30-39.

87. Milliken LA, Going SB, Lohman TG. Effects of variations in regional composition on soft tissue measurements by dual-energy X-ray absorptiometry. *Int J Obes.* 1996;20:677-682.

88. Salamone LM, Fuerst T, Visser M, et al. Measurement of fat mass using DEXA: a validation study in elderly adults. *J Appl Physiol.* 2000;89(1):345-352.

89. Snead DB, Birge SJ, Kohrt WM. Age-related differences in body composition by hydrodensitometry and dual-energy X-ray absorptiometry. *J Appl Physiol.* 1993;74(2):770-775.

90. Valentine RJ, Misic MM, Kessinger RB, Mojtahedi MC, Evans EM. Location of body fat and body size impacts DXA soft tissue measures: a simulation study. *Eur J Clin Nutr.* 2008;62(4):553-559.

91. LaForgia J, Dollman J, Dale MJ, Withers RT, Hill AM. Validation of DXA body composition estimates in obese men and women. *Obesity (Silver Spring).* 2009;17(4):821-826.

92. Wang ZM, Visser M, Ma R, et al. Skeletal muscle mass: evaluation of neutron activation and dual-energy X-ray absorptiometry methods. *J Appl Physiol.* 1996;80(3):824-831.

93. Kim J, Wang Z, Heymsfield SB, Baumgartner RN, Gallagher D. Total-body skeletal muscle mass: estimation by a new dual-energy X-ray absorptiometry method. *Am J Clin Nutr.* 2002;76(2):378-383.

94. Withers RT, Smith DA, Chatterton BE, Schultz CG, Gaffney RD. A comparison of four methods of estimating the body composition of male endurance athletes. *Eur J Clin Nutr.* 1992;46(11):773-784.

95. Prior BM, Cureton KJ, Modlesky CM, et al. In vivo validation of whole body composition estimates from dual-energy X-ray absorptiometry. *J Appl Physiol.* 1997;83(2):623-630.

96. Glickman SG, Marn CS, Supiano MA, Dengel DR. Validity and reliability of dual-energy X-ray absorptiometry for the assessment of abdominal adiposity. *J Appl Physiol.* 2004;97(2):509-514.

97. Taylor AE, Kuper H, Varma RD, et al. Validation of dual energy X-ray absorptiometry measures of abdominal fat by comparison with magnetic resonance imaging in an Indian population. *PLoS One.* 2012;7(12):e51042.

98. Houtkooper LB, Going SB, Sproul J, Blew RM, Lohman TG. Comparison of methods for assessing body-composition changes over 1 y in postmenopausal women. *Am J Clin Nutr.* 2000;72(2):401-406.

99. Tylavsky FA, Lohman TG, Dockrell M, et al. Comparison of the effectiveness of 2 dual-energy X-ray absorptiometers with that of total body water and computed tomography in assessing changes in body composition during weight change. *Am J Clin Nutr.* 2003;77(2):356-363.

100. Nana A, Slater GJ, Stewart AD, Burke LM. Methodology review: using dual-energy X-ray absorptiometry (DXA) for the assessment of body composition in athletes and active people. *Int J Sport Nutr Exerc Metab.* 2013;Epub ahead of print. doi:http://dx.doi.org/10.1123/ijsnem.2013-0228

101. Tylavsky F, Lohman T, Blunt BA, et al. QDR 4500A DXA overestimates fat-free mass compared with criterion methods. *J Appl Physiol.* 2003;94(3):959-965.

102. Nana A, Slater GJ, Hopkins WG, Burke LM. Effects of daily activities on dual-energy X-ray absorptiometry measurements of body composition in active people. *Med Sci Sports Exerc.* 2012;44(1):180-189.

103. Nana A, Slater GJ, Hopkins WG, Burke LM. Effects of exercise sessions on DXA measurements of body composition in active people. *Med Sci Sports Exerc.* 2013;45(1):178-185.

104. Nana A, Slater GJ, Hopkins WG, Burke LM. Techniques for undertaking dual-energy X-ray absorptiometry whole-body scans to estimate body composition in tall and/or broad subjects. *Int J Sport Nutr Exerc Metab.* 2012;22(5):313-322.

105. Silva AM, Heymsfield SB, Sardinha LB. Assessing body composition in taller or broader individuals using dual-energy X-ray absorptiometry: a systematic review. *Eur J Clin Nutr.* 2013;67(10):1012-1021.

106. Hangartner TN, Warner S, Braillon P, Jankowski L, Shepherd J. The official positions of the International Society for Clinical Densitometry: acquisition of dual-energy X-ray absorptiometry body composition and considerations regarding analysis and repeatability of measures. *J Clin Densitom.* 2013;16(4):520-536.

107. Nana A, Slater GJ, Hopkins WG, et al. Importance of standardized DXA protocol for assessing physique changes in athletes. *Int J Sport Nutr Exerc Metab.* 2013;Epub ahead of print. doi:http://dx.doi.org/10.1123/ijsnem.2013-0111

108. Shepherd JA, Baim S, Bilezikian JP, Schousboe JT. Executive summary of the 2013 International Society for Clinical Densitometry Position Development Conference on Body Composition. *J Clin Densitom.* 2013;16(4):489-495.

109. Booth RA, Goddard BA, Paton A. Measurement of fat thickness in man: a comparison of ultrasound, Harpenden calipers and electrical conductivity. *Br J Nutr.* 1966;20(4):719-725.

110. Bullen BA, Quaade F, Olessen E, Lund SA. Ultrasonic reflections used for measuring subcutaneous fat in humans. *Hum Biol.* 1965;37(4):375-384.

111. Hawes SF, Albert A, Healy MJR, Garrow JS. A comparison of soft-tissue radiography, reflected ultrasound, skinfold calipers, and thigh circumference for estimating the thickness of fat overlying the iliac crest and greater trochanter. *Proc Nutr Soc.* 1972;31(3):91A-92A.

112. Müller W, Maughan RJ. The need for a novel approach to measure body composition: is ultrasound an answer? *Br J Sports Med.* 2013;47(16):1001-1002.

113. Pineau JC, Filliard JR, Bocquet M. Ultrasound techniques applied to body fat measurement in male and female athletes. *J Athl Train.* 2009;44(2):142-147.

114. Pineau JC, Guihard-Costa AM, Bocquet M. Validation of ultrasound techniques applied to body fat measurement. A comparison between ultrasound techniques, air displacement plethysmography and bioelectrical impedance vs. dual-energy X-ray absorptiometry. *Ann Nutr Metab.* 2007;51(5):421-427.

115. Stolk RP, Wink O, Zelissen PM, Meijer R, van Gils AP, Grobbee DE. Validity and reproducibility of ultrasonography for the measurement of intra-abdominal adipose tissue. *Int J Obes Relat Metab Disord.* 2001;25(9):1346-1351.

116. Wagner DR. Ultrasound as a tool to assess body fat. *J Obes.* 2013;2013:1-9. doi:http://dx.doi.org/10.1155/2013/280713

117. Weiss LW, Clark FC. The use of B-mode ultrasound for measuring subcutaneous fat thickness on the upper arms. *Res Q Exerc Sport.* 1985;56(1):77-81.

118. Johnson KE, Miller B, Juvancic-Heltzel JA, et al. Agreement between ultrasound and dual-energy X-ray absorptiometry in assessing percentage body fat in college-aged adults. *Clin Physiol Funct Imaging.* 2014;34(6):493-496.

119. Loenneke JP, Barnes JT, Wagganer JD, et al. Validity and reliability of an ultrasound system for estimating adipose tissue. *Clin Physiol Funct Imaging.* 2014;34(2):159-162.

120. Selkow NM, Pietrosimone BG, Saliba SA. Subcutaneous thigh fat assessment: a comparison of skinfold calipers and ultrasound imaging. *J Athl Train.* 2011;46(1):50-54.

121. Smith-Ryan AE, Fultz SN, Melvin MN, Wingfield HL, Woessner MN. Reproducibility and validity of A-mode ultrasound for body composition measurement and classification in overweight and obese men and women. *PLoS One.* 2014;9(3):e91750.

122. Hagen-Ansert SL. *Textbook of Diagnostic Ultrasonography.* 5th ed. St. Louis, MO: Mosby; 2001.

123. Wagner DR, Thompson BJ, Anderson DA, Schwartz S. A-mode and B-mode ultrasound measurement of fat thickness: a cadaver validation study. *Eur J Clin Nutr.* 2018. doi:10.1038/s41430-018-0085-2

124. Toomey C, McCreesh K, Leahy S, Jakeman P. Technical considerations for accurate measurement of subcutaneous adipose tissue thickness using B-mode ultrasound. *Ultrasound.* 2011;19(2):91-96.

125. Bernstein SL, Coble YD Jr, Eisenbrey AB, et al. The future of ultrasonography: report of the ultrasonography task force. *JAMA.* 1991;266:406-409.

126. Gulizia R, Uglietti A, Grisolia A, Gervasoni C, Galli M, Filice C. Proven intra and interobserver reliability in the echographic assessments of body fat changes related to HIV associated Adipose Redistribution Syndrome (HARS). *Curr HIV Res.* 2008;6(4):276-278.

127. Fanelli MT, Kuczmarski RJ. Ultrasound as an approach to assessing body composition. *Am J Clin Nutr.* 1984;39(5):703-709.

128. Pineau JC, Lalys L, Pellegrini M, Battistini NC. Body fat mass assessment: a comparison between an ultrasound-based device and a discovery A model of DXA. *ISRN Obes.* 2013;2013:462394.

129. Horn M, Müller W. Towards an accurate determination of subcutaneous adipose tissue by means of ultrasound. *6th World Congress on Biomechanics.* 2010, Singapore [Poster].

130. Müller W, Horn M, Furhapter-Rieger A, et al. Body composition in sport: a comparison of a novel ultrasound imaging technique to measure subcutaneous fat tissue compared with skinfold measurement. *Br J Sports Med.* 2013;47(16):1028-1035.

131. Müller W, Horn M, Furhapter-Rieger A, et al. Body composition in sport: interobserver reliability of a novel ultrasound measure of subcutaneous fat tissue. *Br J Sports Med.* 2013;47(16):1036-1043.

132. Müller W, Lohman TG, Stewart AD, et al. Subcutaneous fat patterning in athletes: selection of appropriate sites and standardisation of a novel ultrasound measurement technique: ad hoc working group on body composition, health and performance, under the auspices of the IOC Medical Commission. *Br J Sports Med.* 2016;50(1):45-54.

133. Storchle P, Müller W, Sengeis M, et al. Standardized ultrasound measurement of subcutaneous fat patterning: high reliability and accuracy in groups ranging from lean to obese. *Ultrasound Med Biol.* 2017;43(2):427-438.

Chapter 4

1. Lohman TG. *Advances in Human Body Composition.* Champaign, IL: Human Kinetics; 1992.

2. Lohman TG. Skinfolds and body density and their relation to body fatness: a review. *Hum Biol.* 1981;53(2):181-225.

3. Meyer NL, Sundgot-Borgen J, Lohman TG, et al. Body composition for health and performance: a survey of body composition assessment practice carried out by the Ad Hoc Research Working Group on Body Composition, Health and Performance under the auspices of the IOC Medical Commission. *Br J Sports Med.* 2013;47(16):1044-1053.

4. Ackland T, Lohman T, Sundgot-Borgen J, et al. Current status of body composition assessment in sport: review and position statement on behalf of the ad hoc research working group on body composition health and performance, under the auspices of the I.O.C. Medical Commission. *Sports Med.* 2012;42(3):227-249.

5. Lohman TG, Roche AF, Martorell R. *Anthropometric Standardization Reference Manual.* Champaign, IL: Human Kinetics; 1988.

6. Durnin JV, Womersley J. Body fat assessed from total body density and its estimation from skinfold thickness: measurements on 481 men and women aged from 16 to 72 years. *Br J Nutr.* 1974;32(1):77-97.

7. Heyward V, Stolarczyk L. *Applied Body Composition Assessment.* Champaign, IL: Human Kinetics; 1996.

8. Jackson AS, Pollock ML. Practical assessment of body composition. *Phys Sportsmed.* 1985;13(5):76-90.

9. Jackson AS, Pollock ML. Factor analysis and multivariate scaling of anthropometric variables for the assessment of body composition. *Med Sci Sports.* 1976;8(3):196-203.

10. Jackson AS, Pollock ML. Generalized equations for predicting body density of men. *Br J Nutr.* 1978;40(3):497-504.

11. Jackson AS, Pollock ML, Ward A. Generalized equations for predicting body density of women. *Med Sci Sports Exerc.* 1980;12(3):175-181.

12. Peterson MJ, Czerwinski SA, Siervogel RM. Development and validation of skinfold-thickness prediction equations with a 4-compartment model. *Am J Clin Nutr.* 2003;77(5):1186-1191.

13. Evans EM, Rowe DA, Misic MM, Prior BM, Arngrimsson SA. Skinfold prediction equation for athletes developed using a four-component model. *Med Sci Sports Exerc.* 2005;37(11):2006-2011.

14. Thorland WG, Tipton CM, Lohman TG, et al. Midwest wrestling study: prediction of minimal weight for high school wrestlers. *Med Sci Sports Exerc.* 1991;23(9):1102-1110.

15. Slaughter MH, Lohman TG, Boileau RA, et al. Skinfold equations for estimation of body fatness in children and youth. *Hum Biol.* 1988;60(5):709-723.

16. Stevens J, Cai J, Truesdale KP, Cuttler L, Robinson TN, Roberts AL. Percent body fat prediction equations for 8- to 17-year-old American children. *Pediatr Obes.* 2014;9(4):260-271.

17. Stevens J, Ou FS, Cai J, Heymsfield SB, Truesdale KP. Prediction of percent body fat measurements in Americans 8 years and older. *Int J Obes (Lond).* 2016;40(4):587-594.

18. Jackson AS, Ellis KJ, McFarlin BK, Sailors MH, Bray MS. Cross-validation of generalised body composition equations with diverse young men and women: the Training Intervention and Genetics of Exercise Response (TIGER) Study. *Br J Nutr.* 2009;101(6):871-878.

19. O'Connor DP, Bray MS, McFarlin BK, Sailors MH, Ellis KJ, Jackson AS. Generalized equations for estimating DXA percent fat of diverse young women and men: the TIGER study. *Med Sci Sports Exerc.* 2010;42(10):1959-1965.

20. Davidson LE, Wang J, Thornton JC, et al. Predicting fat percent by skinfolds in racial groups: Durnin and Womersley revisited. *Med Sci Sports Exerc.* 2011;43(3):542-549.

21. Schoeller DA, Tylavsky FA, Baer DJ, et al. QDR 4500A dual-energy X-ray absorptiometer underestimates fat mass in comparison with criterion methods in adults. *Am J Clin Nutr.* 2005;81(5):1018-1025.

22. Brozek J. Body measurements, including skinfold thickness, as indicators of body composition. In: Brozek J, Henschel A, eds. *Techniques for Measuring Body Composition.* Washington, DC: National Academy of Science; 1963: 3-35.

23. Lohman T, Roche A, Martorell R. *Anthropometric Standardization Reference Manual.* Champaign, IL: Human Kinetics; 1991.

24. Pascale LR, Grossman MI, Sloane HS, Frankel T. Correlations between thickness of skin-folds and body density in 88 soldiers. *Hum Biol.* 1956;28(2):165-176.

25. Pollack ML, Schmidt DH, Jackson AS. Measurement of cardio-respiratory fitness and body composition in the clinical setting. *Compr Ther.* 1980;6(9):12-27.

26. Heyward VH. Practical body composition assessment for children, adults, and older adults. *Int J Sport Nutr.* 1998;8(3):285-307.

27. Wang J, Thornton JC, Kolesnik S, Pierson RN. Anthropometry in body composition: an overview. *Ann N Y Acad Sci.* 2000;904:317-326.

28. McArdle WD, Katch FI, Katch VL. *Essentials of Exercise Physiology.* 3rd ed. Baltimore, MD: Lippincott Williams & Wilkins; 2006.

29. Seibert H, Pereira AM, Ajzen SA, Nogueira PC. Abdominal circumference measurement by ultrasound does not enhance estimating the association of visceral fat with cardiovascular risk. *Nutrition.* 2013;29(2):393-398.

30. Taylor HA Jr, Coady SA, Levy D, et al. Relationships of BMI to cardiovascular risk factors differ by ethnicity. *Obesity (Silver Spring).* 2010;18(8):1638-1645.

31. Wang J, Thornton JC, Bari S, et al. Comparisons of waist circumferences measured at 4 sites. *Am J Clin Nutr.* 2003;77(2):379-384.

32. Elliott WL. Criterion validity of a computer-based tutorial for teaching waist circumference self-measurement. *J Bodyw Mov Ther.* 2008;12(2):133-145.

33. Carranza Leon BG, Jensen MD, Hartman JJ, Jensen TB. Self-measured vs professionally measured waist circumference. *Ann Fam Med.* 2016;14(3):262-266.

34. Brambilla P, Bedogni G, Moreno LA, et al. Crossvalidation of anthropometry against magnetic resonance imaging for the assessment of visceral and subcutaneous adipose tissue in children. *Int J Obes (Lond).* 2006;30(1): 23-30.

35. Bouchard C. BMI, fat mass, abdominal adiposity and visceral fat: where is the 'beef'? *Int J Obes (Lond).* 2007;31(10):1552-1553.

36. Friedl KE, Westphal KA, Marchitelli LJ, Patton JF, Chumlea WC, Guo SS. Evaluation of anthropometric equations to assess body-composition changes in young women. *Am J Clin Nutr.* 2001;73(2):268-275.

37. Lyra CO, Lima SC, Lima KC, Arrais RF, Pedrosa LF. Prediction equations for fat and fat-free body mass in adolescents, based on body circumferences. *Ann Hum Biol.* 2012;39(4):275-280.

38. Cameron AJ, Magliano DJ, Soderberg S. A systematic review of the impact of including both waist and hip circumference in risk models for cardiovascular diseases, diabetes and mortality. *Obes Rev.* 2013;14(1):86-94.

39. Goh LG, Dhaliwal SS, Welborn TA, Lee AH, Della PR. Anthropometric measurements of general and central obesity and the prediction of cardiovascular disease risk in women: a cross-sectional study. *BMJ Open.* 2014;4(2):e004138.

40. Esteghamati A, Mousavizadeh M, Noshad S, Shoar S, Khalilzadeh O, Nakhjavani M. Accuracy of anthropometric parameters in identification of high-risk patients predicted with cardiovascular risk models. *Am J Med Sci.* 2013;346(1):26-31.

41. Mohebi R, Bozorgmanesh M, Azizi F, Hadaegh F. Effects of obesity on the impact of short-term changes in anthropometric measurements on coronary heart disease in women. *Mayo Clin Proc.* 2013;88(5):487-494.

42. Rodrigues SL, Baldo MP, Lani L, Nogueira L, Mill JG, Sa Cunha R. Body mass index is not independently associated with increased aortic stiffness in a Brazilian population. *Am J Hypertens.* 2012;25(10):1064-1069.

43. Seidell JC, Perusse L, Despres JP, Bouchard C. Waist and hip circumferences have independent

and opposite effects on cardiovascular disease risk factors: the Quebec Family Study. *Am J Clin Nutr.* 2001;74(3):315-321.

44. Tresignie J, Scafoglieri A, Pieter Clarys J, Cattrysse E. Reliability of standard circumferences in domain-related constitutional applications. *Am J Hum Biol.* 2013;25(5):637-642.

45. Katch FI, McArdle WD. Prediction of body density from simple anthropometric measurements in college-age men and women. *Hum Biol.* 1973;45(3):445-455.

46. Hodgdon JA, Beckett MB. Prediction of percent body fat for U.S. Navy men from body circumferences and height. Report No. 84-11. 1984. www.dtic.mil/docs/citations/ADA143890

47. Hodgdon JA, Beckett MB. Prediction of percent body fat for U.S. Navy women from body circumferences and height. Report No. 84-29. 1984. www.dtic.mil/docs/citations/ADA146456

48. Schuna JM Jr, Hilgers SJ, Manikowske TL, Tucker JM, Liguori G. The evaluation of a circumference-based prediction equation to assess body composition changes in men. *Int J Exerc Sci.* 2013;6(3):188-198.

49. Garcia AL, Wagner K, Hothorn T, Koebnick C, Zunft HJ, Trippo U. Improved prediction of body fat by measuring skinfold thickness, circumferences, and bone breadths. *Obes Res.* 2005;13(3):626-634.

50. Van Der Ploeg GE, Withers RT, Laforgia J. Percent body fat via DEXA: comparison with a four-compartment model. *J Appl Physiol.* 2003;94(2):499-506.

51. Lukaski HC, Bolonchuk WW, Hall CB, Siders WA. Validation of tetrapolar bioelectrical impedance method to assess human body composition. *J Appl Physiol.* 1986;60(4):1327-1332.

52. Sun SS, Chumlea WC, Heymsfield SB, et al. Development of bioelectrical impedance analysis prediction equations for body composition with the use of a multicomponent model for use in epidemiologic surveys. *Am J Clin Nutr.* 2003;77(2):331-340.

53. Kushner RF, Schoeller DA. Estimation of total body water by bioelectrical impedance analysis. *Am J Clin Nutr.* 1986;44:417-424.

54. Davies PS, Jagger SE, Reilly JJ. A relationship between bioelectrical impedance and total body water in young adults. *Ann Hum Biol.* 1990;17(5):445-448.

55. Schols AM, Wouters EF, Soeters PB, Westerterp KR. Body composition by bioelectrical-impedance analysis compared with deuterium dilution and skinfold anthropometry in patients with chronic obstructive pulmonary disease. *Am J Clin Nutr.* 1991;53(2):421-424.

56. Van Loan MD, Mayclin PL. Bioelectrical impedance analysis: is it a reliable estimator of lean body mass and total body water? *Hum Biol.* 1987;59:299-309.

57. Van Loan MD, Boileau RA, Slaughter MH, et al. Association of bioelectrical resistance with estimates of fat-free mass determined by densitometry and hydrometry. *Am J Hum Biol.* 1990;2:219-226.

58. Van Loan MD, Withers P, Matthie J, Mayclin PL. *Use of Bioimpedance Spectroscopy to Determine Extracellular Fluid, Intracellular Fluid, Total Body Water, and Fat-Free Mass.* Vol. 60. New York: Plenum Publishing; 1993.

59. Nunez C, Gallagher D, Visser M, Pi-Sunyer FX, Wang Z, Heymsfield SB. Bioimpedance analysis: evaluation of leg-to-leg system based on pressure contact footpad electrodes. *Med Sci Sports Exerc.* 1997;29(4):524-531.

60. Bracco D, Thiebaud D, Chiolero RL, Landry M, Burckhardt P, Schutz Y. Segmental body composition assessed by bioelectrical impedance analysis and DEXA in humans. *J Appl Physiol.* 1996;81(6):2580-2587.

61. Rockamann RA, Dalton EK, Arabas JL, Jorn L, Mayhew JL. Validity of arm-to-arm BIA devices compared to DXA for estimating % fat in college men and women. *Int J Exerc Sci.* 2017;10(7):977-988.

62. Bosy-Westphal A, Schautz B, Later W, Kehayias J, Gallagher D, Müller MJ. What makes a BIA equation unique? Validity of eight-electrode multifrequency BIA to estimate body composition in a healthy adult population. *Eur J Clin Nutr.* 2013;67(Suppl 1):S14-S21.

63. Bosy-Westphal A, Later W, Hitze B, et al. Accuracy of bioelectrical impedance consumer devices for measurement of body composition in comparison to whole body magnetic resonance imaging and dual X-ray absorptiometry. *Obes Facts.* 2008;1(6):319-324.

64. Kim M, Kim H. Accuracy of segmental multi-frequency bioelectrical impedance analysis for assessing whole-body and appendicular fat mass and lean soft tissue mass in frail women aged 75 years and older. *Eur J Clin Nutr.* 2013;67(4):395-400.

65. Kriemler S, Puder J, Zahner L, Roth R, Braun-Fahrlander C, Bedogni G. Cross-validation of bioelectrical impedance analysis for the assessment of body composition in a representative sample of 6- to 13-year-old children. *Eur J Clin Nutr.* 2009;63(5):619-626.

66. Jaffrin MY, Morel H. Body fluid volumes measurements by impedance: a review of bioimpedance spectroscopy (BIS) and bioimpedance analysis (BIA) methods. *Med Eng Phys.* 2008;30(10):1257-1269.

67. De Lorenzo A, Andreoli A, Matthie J, Withers P. Predicting body cell mass with bioimpedance by using theoretical methods: a technological review. *J Appl Physiol.* 1997;82(5):1542-1558.

68. Kyle UG, Bosaeus I, De Lorenzo AD, et al. Bioelectrical impedance analysis—part I: review of principles and methods. *Clin Nutr.* 2004;23(5):1226-1243.

69. Kyle UG, Bosaeus I, De Lorenzo AD, et al. Bioelectrical impedance analysis—part II: utilization in clinical practice. *Clin Nutr.* 2004;23(6):1430-1453.

70. National Institutes of Health Technology Assessment Conference Statement. Bioelectrical impedance analysis in body composition measurement. December 12-14, 1994. *Nutrition.* 1996;12(11-12):749-762.

71. Lohman T, Thompson J, Going S, et al. Indices of changes in adiposity in American Indian children. *Prev Med.* 2003;37(Suppl 1):S91-S96.

72. Elia M. Body composition by whole-body bioelectrical impedance and prediction of clinically relevant outcomes: overvalued or underused? *Eur J Clin Nutr.* 2013;67(Suppl 1):S60-S70.

73. Lutoslawska G, Malara M, Tomaszewski P, et al. Relationship between the percentage of body fat and surrogate indices of fatness in male and female Polish active and sedentary students. *J Physiol Anthropol.* 2014;33:10.

74. Santos DA, Matias CN, Rocha PM, et al. Association of basketball season with body composition in elite junior players. *J Sports Med Phys Fitness.* 2014;54(2):162-173.

75. Bergman RN, Stefanovski D, Buchanan TA, et al. A better index of body adiposity. *Obesity (Silver Spring).* 2011;19(5):1083-1089.

76. Krakauer NY, Krakauer JC. A new body shape index predicts mortality hazard independently of body mass index. *PLoS One.* 2012;7(7):e39504.

77. Thomas DM, Bredlau C, Bosy-Westphal A, et al. Relationships between body roundness with body fat and visceral adipose tissue emerging from a new geometrical model. *Obesity (Silver Spring).* 2013;21(11):2264-2271.

78. Assessing your weight. Centers for Disease Control and Prevention website. www.cdc.gov/healthyweight/assessing/. Updated May 15, 2015. Accessed December 17, 2015.

79. Mooney SJ, Baecker A, Rundle AG. Comparison of anthropometric and body composition measures as predictors of components of the metabolic syndrome in a clinical setting. *Obes Res Clin Pract.* 2013;7(1):e55-e66.

80. Suchanek P, Kralova Lesna I, Mengerova O, Mrazkova J, Lanska V, Stavek P. Which index best correlates with body fat mass: BAI, BMI, waist or WHR? *Neuro Endocrinol Lett.* 2012;33(Suppl 2):78-82.

81. World Health Organization. Waist circumference and waist-hip ratio: report of a WHO expert consultation. Geneva, December 8-11, 2008. Technical Report, World Health Organization. 2011.

82. Freedman DS, Blanck HM, Dietz WH. Is the body adiposity index (hip circumference/height$^{1.5}$) more strongly related to skinfold thicknesses and risk factor levels than is BMI? The Bogalusa Heart Study. *Br J Nutr.* 2012;108(11):2100-2106.

83. Freedman DS, Thornton JC, Pi-Sunyer FX, et al. The body adiposity index (hip circumference ÷ height$^{1.5}$) is not a more accurate measure of adiposity than is BMI, waist circumference, or hip circumference. *Obesity (Silver Spring).* 2012;20(12):2438-2444.

84. Martin BJ, Verma S, Charbonneau F, Title LM, Lonn EM, Anderson TJ. The relationship between anthropometric indexes of adiposity and vascular function in the FATE cohort. *Obesity (Silver Spring).* 2013;21(2):266-273.

85. Tian S, Zhang X, Xu Y, Dong H. Feasibility of body roundness index for identifying a clustering of cardiometabolic abnormalities compared to BMI, waist circumference and other anthropometric indices: the China Health and Nutrition Survey, 2008 to 2009. *Medicine (Baltimore).* 2016;95(34):e4642.

86. Thomas TR, Londeree BR, Lawson DA, Kolkhorst FW. Resting metabolic rate before exercise vs a control day. *Am J Clin Nutr.* 1994;59:28-31.

87. Chang Y, Guo X, Chen Y, et al. A body shape index and body roundness index: two new body indices to identify diabetes mellitus among rural populations in northeast China. *BMC Public Health.* 2015;15(1):794.

88. Santos DA, Silva AM, Matias CN, et al. Utility of novel body indices in predicting fat mass in elite athletes. *Nutrition.* 2015;31(7-8):948-954.

Chapter 5

1. Lohman TG, Roche AF, Martorell R. *Anthropometric Standardization Reference Manual.* Champaign, IL: Human Kinetics; 1988.

2. Heymsfield S, Lohman TG, Wang ZM, Going SB, eds. *Human Body Composition.* 2nd ed. Champaign, IL: Human Kinetics; 2005.

3. Guo SS, Chumlea, WC. Statistical methods for the development and testing of predictive equations. In: Roche AF, Heymsfield SB, Lohman TG, eds. *Human Body Composition.* Champaign, IL: Human Kinetics; 1996:191-202.

4. Ulijaszek SJ, Kerr DA. Anthropometric measurement error and the assessment of nutritional status. *Br J Nutr.* 1999;82(3):165-177.

5. Friedl KE, DeLuca JP, Marchitelli LJ, Vogel JA. Reliability of body-fat estimations from a four-component model by using density, body water, and bone mineral measurements. *Am J Clin Nutr.* 1992;55:764-770.

6. Heymsfield SB, Waki M. Body composition in humans: advances in the development of multicompartment chemical models. *Nutr Rev.* 1991;49:97-108.

7. Wells JCK, Fuller NJ, Dewit O, Fewtrell MS, Elia M, Cole TJ. Four-component model of body composition in children: density and hydration of fat-free mass and comparison with simpler models. *Am J Clin Nutr.* 1999;69:904-912.

8. Withers RT, LaForgia J, Pillans RK, et al. Comparisons of two-, three-, and four-compartment models of body composition analysis in men and women. *J Appl Physiol.* 1998;85(1):238-245.

9. Heymsfield SB, Lichtman S, Baumgartner RN, et al. Body composition of humans: comparison of two improved four-compartment models that differ in expense, technical complexity, and radiation exposure. *Am J Clin Nutr.* 1990;52(1):52-58.

10. Akers R, Buskirk ER. An underwater weighing system utilizing "force cube" transducers. *J Appl Physiol.* 1969;26(5):649-652.

11. Van der Ploeg G, Gunn S, Withers R, Modra A, Crockett A. Comparison of two hydrodensitometric methods for estimating percent body fat. *J Appl Phsyiol.* 2000;88:1175-1180.

12. Fields DA, Goran MI, McCrory MA. Body-composition assessment via air-displacement plethysmography in adults and children: a review. *Am J Clin Nutr.* 2002;75(3):453-467.

13. McCrory MA, Gomez TD, Bernauer EM, Molé PA. Evaluation of a new air displacement plethysmograph for measuring human body composition. *Med Sci Sports Exerc.* 1995;27(12):1686-1691.

14. Schoeller DA, Van Santen E, Peterson WM, Dietz W, Jaspan J, Klein PD. Total body water measurement in humans with ^{18}O and ^{2}H labeled water. *Am J Clin Nutr.* 1980;33:2686-2693.

15. Toombs RJ, Ducher G, Shepherd JA, De Souza MJ. The impact of recent technological advances

on the trueness and precision of DXA to assess body composition. *Obesity (Silver Spring)*. 2012;20(1):30-39.

16. Gutin B, Litaker M, Islam S, Manos T, Smith C, Treiber F. Body-composition measurement in 9-11-y-old children by dual-energy X-ray absorptiometry, skinfold-thickness measurements, and bioimpedance analysis. *Am J Clin Nutr*. 1996;63(3):287-292.

17. Fuller NJ, Laskey MA, Elia M. Assessment of the composition of major body regions by dual-energy X-ray absorptiometry (DEXA), with special reference to limb muscle mass. *Clin Physiol*. 1992;12(3):253-266.

18. Hind K, Oldroyd B, Truscott JG. In vivo precision of the GE Lunar iDXA densitometer for the measurement of total body composition and fat distribution in adults. *Eur J Clin Nutr*. 2011;65(1):140-142.

19. Vicente-Rodriguez G, Rey-Lopez JP, Mesana MI, et al. Reliability and intermethod agreement for body fat assessment among two field and two laboratory methods in adolescents. *Obesity (Silver Spring)*. 2012;20(1):221-228.

20. Gore C, Norton K, Olds T, et al. Accreditation in anthropometry: an Australian model. Chapter 13. In: Norton K, Olds T, eds. *Anthropometrica*. Sydney, Australia: UNSW Press; 1996:395-411.

21. Stomfai S, Ahrens W, Bammann K, et al. Intra- and inter-observer reliability in anthropometric measurements in children. *Int J Obes (Lond)*. 2011;35(Suppl 1):S45-S51.

22. Nagy E, Vicente-Rodriguez G, Manios Y, et al. Harmonization process and reliability assessment of anthropometric measurements in a multicenter study in adolescents. *Int J Obes (Lond)*. 2008;32(Suppl 5):S58-S65.

23. Harrison GG, Buskirk ER, Carter JL, et al. Skinfold thicknesses and measurement technique. In: Lohman TG, Roche AF, Martorell R, eds. *Anthropometric standardization reference manual*. Champaign, IL: Human Kinetics; 1988:177.

24. Wang J, Thornton JC, Bari S, et al. Comparisons of waist circumferences measured at 4 sites. *Am J Clin Nutr*. 2003;77(2):379-384.

25. Callaway CW. Circumferences. In: Lohman TG, Roche AF, Martorell R, eds. *Anthropometric Standardization Reference Manual*. Champaign, IL: Human Kinetics; 1988:39-54.

26. Verweij LM, et al. Measurement error of waist circumference: gaps in knowledge. *Public Health Nutr*. 2013;16(2):281-288.

27. Schaefer F, Georgi M, Zieger A, Scharer K. Usefulness of bioelectric impedance and skinfold measurements in predicting fat-free mass derived from total body potassium in children. *Pediatr Res*. 1994;35(5):617-624.

28. Lohman TG. *Advances in Human Body Composition*. Champaign, IL: Human Kinetics; 1992.

29. Crespi CM, Alfonso VH, Whaley SE, Wang MC. Validity of child anthropometric measurements in the Special Supplemental Nutrition Program for Women, Infants, and Children. *Pediatr Res*. 2012;71(3):286-292.

30. Lohman TG. Research progress in validation of laboratory methods of assessing body composition. *Med Sci Sports Exerc*. 1984;16(6):596-605.

31. Accreditation Scheme. The International Society for the Advancement of Kinanthropometry website. https://www.isak.global/FormationSystem/ AccreditationScheme. Accessed April 20, 2018.

Chapter 6

1. Oppliger RA, Case HS, Horswill CA, Landry GL, Shelter AC. American College of Sports Medicine position stand: weight loss in wrestlers. *Med Sci Sports Exerc*. 1996;28(6):ix-xii.

2. Sinning WE. Body composition in athletes. In: Roche AF, Heymsfield SB, Lohman TG, eds. *Human Body Composition*. Champaign, IL: Human Kinetics; 1996:257-273.

3. National Collegiate Athletic Administration (NCAA). Memorandum: NCAA Wrestling Weight Management Program for 2010-11. August 20, 2010. http://fs.ncaa.org/Docs/rules/ wrestling/2010/WM_preseason_mailing.pdf. Accessed May 1, 2018.

4. National Federation of State High School Associations. Wrestling rule book. Elgin, IL: National Federation of State High School Associations.

5. Lohman TG. *Advances in Body Composition Assessment.* Champaign, IL: Human Kinetics; 1992.

6. Behnke AR. New concepts of height-weight relationships. In: Wilson NL, ed. *Obesity.* Philadelphia, PA: FA Davis; 1969:25-35.

7. Behnke AR, Wilmore JH. *Evaluation and regulation of body build and composition.* Englewood Cliffs, NJ: Prentice Hall; 1974.

8. Katch VL, Campaigne B, Freedson P, Sady S, Katch FI, Behnke AR. Contribution of breast volume and weight to body fat distribution in females. *Am J Phys Anthropol.* 1980;53(1):93-100.

9. Klungland Torstveit M, Sundgot-Borgen J. Are under- and overweight female elite athletes thin and fat? A controlled study. *Med Sci Sports Exerc.* 2012;44(5):949-957.

10. Sundgot-Borgen J, Meyer NL, Lohman TG, et al. How to minimise risks for weight-sensitive sports review and position statement on behalf of the Ad Hoc Research Working Group on Body Composition, Health and Performance, under the auspices of the IOC Medical Commission. *Br J Sports Med.* 2013;47:1012-1022.

11. Thorland WG, Tipton CM, Lohman TG, et al. Midwest wrestling study: prediction of minimal weight for high school wrestlers. *Med Sci Sports Exerc.* 1991;23(9):1102-1110.

12. Clark RR, Sullivan JC, Bartok CJ, Carrel AL. DXA provides a valid minimum weight in wrestlers. *Med Sci Sports Exerc.* 2007;39(11):2069-2075.

13. Clark RR, Bartok C, Sullivan JC, Schoeller DA. Minimum weight prediction methods cross-validated by the four-component model. *Med Sci Sports Exerc.* 2004;36(4):639-647.

14. Sinning WE. Body composition assessment of college wrestlers. *Med Sci Sports.* 1974;6(2):139-145.

15. Clark RR, Sullivan JC, Bartok C, Schoeller DA. Multicomponent cross-validation of minimum weight predictions for college wrestlers. *Med Sci Sports Exerc.* 2003;35(2):342-347.

16. Lohman TG. Applicability of body composition techniques and constants for children and youths. In: Pandolf KB, ed. *Exercise and Sport Sciences Reviews.* Vol. 14. New York: Macmillan Publishing Co.; 1986:325-357.

17. Lohman TG. Skinfolds and body density and their relation to body fatness: a review. *Hum Biol.* 1981;53:181-225.

18. Siri WE. Body composition from fluid spaces and density: analysis of methods. In: Brozek J, Henschel A, eds. *Techniques for Measuring Body Composition.* National Academy of Sciences: Washington, DC; 1961:223-244.

19. Visser M, Gallagher D, Deurenberg P, Wang J, Pierson RN, Heymsfield SB. Density of fat-free body mass: relationship with race, age and level of body fatness. *Am J Physiol Endocrinol Metab.* 1997;272(35):E781-E787.

20. Modlesky CM, Cureton KJ, Lewis RD, Prior BM, Sloniger MA, Rowe DA. Density of the fat-free mass and estimates of body composition in male weight trainers. *J Appl Physiol.* 1996;80(6):2085-2096.

21. Lohman TG. *Advances in Human Body Composition.* Champaign, IL: Human Kinetics; 1992.

22. Going SB, Massett MP, Hall MC, et al. Detection of small changes in body composition by dual-energy X-ray absorptiometry. *Am J Clin Nutr.* 1993;57:845-850.

23. Pietrobelli A, Wang Z, Formica C, Heymsfield SB. Dual-energy X-ray absorptiometry: fat estimation errors due to variation in soft tissue hydration. 1998;274(5):E808-E816.

24. Ackland TR, Lohman TG, Sundgot-Borgen J, et al. Current status of body composition assessment in sport. *Sports Med.* 2012;42(3):227-249.

25. Toombs RJ, Ducher G, Shepherd JA, De Souza MJ. The impact of recent technological advances on the trueness and precision of DXA to assess body composition. *Obesity (Silver Spring).* 2012;20(1):30-39.

26. Evans EM, Prior BM, Modlesky CM. A mathematical method to estimate body composition in tall individuals using DXA. *Med Sci Sports Exerc.* 2005;37(7):1211-1215.

27. Nana A, Slater GJ, Hopkins WG, Burke LM. Techniques for undertaking dual-energy X-ray absorptiometry whole-body scans to estimate body composition in tall and/or broad subjects. *Int J Sport Nutr Exerc Metab.* 2012;22:313-322.

28. Schoeller DA, Tylavsky FA, Baer DJ, et al. QDR 4500A dual-energy X-ray absorptiometer underestimates fat mass in comparison with criterion methods in adults. *Am J Clin Nutr.* 2005;81(5):1018-1025.

29. Santos DA, Silva AM, Matias CN, Fields DA, Heymsfield SB, Sardinha LB. Accuracy of DXA in estimating body composition changes in elite athletes using a four compartment model as the reference method. *Nutr Metab.* 2010;7:22.

30. Arngrimsson SA, Evans EM, Saunders MJ, Ogburn CL III, Lewis RD, Cureton KJ. Validation of body composition estimates in male and female distance runners using estimates from a four-component model. *Am J Hum Biol.* 2000;12(3):301-314.

31. Withers RT, LaForgia J, Pillans RK, et al. Comparisons of two-, three-, and four-compartment models of body composition analysis in men and women. *J Appl Physiol.* 1998;85(1):238-245.

32. Prior BM, Cureton KJ, Modlesky CM, et al. In vivo validation of whole body composition estimates from dual energy X-ray absorptiometry. *J Appl Physiol.* 1997;83(2):623-630.

33. Schoeller DA, Roche AF, Heymsfield SB, Lohman TG. Hydrometry. In: Roche AF, Heymsfield SB, Lohman TG. *Human Body Composition.* Champaign: Human Kinetics; 1996:25-44.

34. Evans EM, Rowe DA, Misic MM, Prior BM, Arngrimsson SA. Skinfold prediction equation for athletes developed using a four-component model. *Med Sci Sports Exerc.* 2005;37(11):2006-2011.

35. Thorland WG, Johnson GO, Tharp GD, Fagot TG, Hammer RW. Validity of anthropometric equations for the estimation of body density in adolescent athletes. *Med Sci Sports Exerc.* 1984;16(1):77-81.

36. Müller W, Horn M, Furhapter-Rieger A, et al. Body composition in sport: interobserver reliability of a novel ultrasound measure of subcutaneous fat tissue. *Br J Sports Med.* 2013;47(16):1036-1043.

37. Müller W, Lohman TG, Stewart AD, et al. Subcutaneous fat patterning in athletes: selection of appropriate sites and standardisation of a novel ultrasound measurement technique: ad hoc working group on body composition, health and performance, under the auspices of the IOC Medical Commission. *Br J Sports Med.* 2016;50(1):45-54.

38. Storchle P, Müller W, Sengeis M, et al. Standardized ultrasound measurement of subcutaneous fat patterning: high reliability and accuracy in groups ranging from lean to obese. *Ultrasound Med Biol.* 2017;43(2):427-438.

39. Sinning WE, Wilson JR. Validation of "generalized" equations for body composition analysis in women athletes. *Res Q Exerc Sport.* 1984;55:153-160.

40. Sinning WE, Dolny DG, Little KD, et al. Validity of "generalized" equations for body composition analysis in male athletes. *Med Sci Sports Exerc.* 1985;17(1):124-130.

41. Behnke AR. The estimation of lean body weight from "skeletal" measurements. *Hum Biol.* 1959;31:295-315.

42. Tcheng TK, Tipton CM. Iowa wrestling study: anthropometric measurements and the prediction of a "minimal" body weight for high school wrestlers. *Med Sci Sports.* 1973;5(1):1-10.

43. Lohman TG. Skinfolds and body density and their relation to body fatness: a review. *Hum Biol.* 1981;53(2):181-225.

44. Sloan AW. Estimation of body fat in young men. *J Appl Physiol.* 1967;23(3):311-315.

45. Jackson AS, Pollock ML, Ward A. Generalized equations for predicting body density of women. *Med Sci Sports Exerc.* 1980;12(3):175-181.

46. Utter AC, Lambeth PG. Evaluation of multifrequency bioelectrical impedance analysis in assessing body composition of wrestlers. *Med Sci Sports Exerc.* 2010;42(2):361-367.

47. Utter AC, Scott JR, Oppliger RA, et al. A comparison of leg-to-leg bioelectrical impedance

and skinfolds in assessing body fat in collegiate wrestlers. *J Strength Cond Res.* 2001;15(2):157-160.

48. Moon JR. Body composition in athletes and sports nutrition: an examination of the bioimpedance analysis technique. *Eur J Clin Nutr.* 2013;67(Suppl 1):S54-S59.

49. Lukaski HC, Bolonchuk WW, Hall CB, Siders WA. Validation of tetrapolar bioelectrical impedance method to assess human body composition. *J Appl Physiol.* 1986;60(4):1327-1332.

50. Bartok C, Schoeller DA, Randall Clark R, Sullivan JC, Landry GL. The effect of dehydration on wrestling minimum weight assessment. *Med Sci Sports Exerc.* 2004;36(1):160-167.

51. Clark RR, Bartok C, Sullivan JC, Schoeller DA. Is leg-to-leg BIA valid for predicting minimum weight in wrestlers? *Med Sci Sports Exerc.* 2005;37(6):1061-1068.

52. Hetzler RK, Kimura IF, Haines K, Labotz M, Smith J. A comparison of bioelectrical impedance and skinfold measurements in determining minimum wrestling weights in high school wrestlers. *J Athl Train.* 2006;41(1):46-51.

Chapter 7

1. Visser M, Gallagher D, Deurenberg P, Wang J, Pierson RN Jr., Heymsfield SB. Density of fat-free body mass: relationship with race, age, and level of body fatness. *Am J Physiol.* 1997;272(5 Pt 1):E781-E787.

2. Roemmich JN, Clark PA, Weltman A, Rogol AD. Alterations in growth and body composition during puberty. I. Comparing multicompartment body composition models. *J Appl Physiol.* 1997;83(3):927-935.

3. Bemben MG, Massey BH, Bemben DA, Boileau RA, Misner JE. Age-related variability in body composition methods for assessment of percent fat and fat-free mass in men aged 20-74 years. *Age Ageing.* 1998;27(2):147-153.

4. Streat SJ, Beddoe AH, Hill GL. Measurement of body fat and hydration of the fat-free body in health and disease. *Metabolism.* 1985;34(6):509-518.

5. Heyward VH, Wagner DR. *Applied Body Composition Assessment.* 2nd ed. Champaign, IL: Human Kinetics; 2004.

6. Pourhassan M, Schautz B, Braun W, Gluer CC, Bosy-Westphal A, Müller MJ. Impact of body-composition methodology on the composition of weight loss and weight gain. *Eur J Clin Nutr.* 2013;67(5):446-454.

7. Lee SY, Gallagher D. Assessment methods in human body composition. *Curr Opin Clin Nutr Metab Care.* 2008;11(5):566-572.

8. Modlesky CM, Cureton KJ, Lewis RD, Prior BM, Sloniger MA, Rowe DA. Density of the fat-free mass and estimates of body composition in male weight trainers. *J Appl Physiol.* 1996;80(6):2085-2096.

9. Prior BM, Modlesky CM, Evans EM, et al. Muscularity and the density of the fat-free mass in athletes. *J Appl Physiol.* 2001;90(4):1523-1531.

10. Wells JC, Williams JE, Chomtho S, et al. Pediatric reference data for lean tissue properties: density and hydration from age 5 to 20 y. *Am J Clin Nutr.* 2010;91(3):610-618.

11. Lohman TG, Hingle M, Going SB. Body composition in children. *Pediatr Exerc Sci.* 2013;25(4):573-590.

12. Silva AM, Fields DA, Quiterio AL, Sardinha LB. Are skinfold-based models accurate and suitable for assessing changes in body composition in highly trained athletes? *J Strength Cond Res.* 2009;23(6):1688-1696.

13. van der Ploeg GE, Brooks AG, Withers RT, Dollman J, Leaney F, Chatterton BE. Body composition changes in female bodybuilders during preparation for competition. *Eur J Clin Nutr.* 2001;55(4):268-277.

14. Withers RT, Noell CJ, Whittingham NO, Chatterton BE, Schultz CG, Keeves JP. Body composition changes in elite male bodybuilders during preparation for competition. *Aust J Sci Med Sport.* 1997;29(1):11-16.

15. Santos DA, Matias CN, Rocha PM, et al. Association of basketball season with body composition in elite junior players. *J Sports Med Phys Fitness.* 2014;54(2):162-173.

16. Hewitt MJ, Going SB, Williams DP, Lohman TG. Hydration of the fat-free body mass in children and adults: implications for body composition assessment. *Am J Physiol*. 1993;265(1 Pt 1):E88-E95.

17. Fogelholm GM, Sievanen HT, van Marken Lichtenbelt WD, Westerterp KR. Assessment of fat-mass loss during weight reduction in obese women. *Metabolism*. 1997;46(8):968-975.

18. Duren DL, Sherwood RJ, Czerwinski SA, et al. Body composition methods: comparisons and interpretation. *J Diabetes Sci Technol*. 2008;2(6):1139-1146.

19. Haroun D, Wells JC, Williams JE, Fuller NJ, Fewtrell MS, Lawson MS. Composition of the fat-free mass in obese and nonobese children: matched case-control analyses. *Int J Obes*. 2005;29(1):29-36.

20. Lof M, Forsum E. Hydration of fat-free mass in healthy women with special reference to the effect of pregnancy. *Am J Clin Nutr*. 2004;80(4):960-965.

21. Nana A, Slater GJ, Hopkins WG, Burke LM. Effects of exercise sessions on DXA measurements of body composition in active people. *Med Sci Sports Exerc*. 2013;45(1):178-185.

22. Nana A, Slater GJ, Hopkins WG, Burke LM. Effects of daily activities on dual-energy X-ray absorptiometry measurements of body composition in active people. *Med Sci Sports Exerc*. 2012;44(1):180-189.

23. Nana A, Slater GJ, Hopkins WG, Burke LM. Techniques for undertaking dual-energy X-ray absorptiometry whole-body scans to estimate body composition in tall and/or broad subjects. *Int J Sport Nutr Exerc Metabol*. 2012;22(5):313-322.

24. LaForgia J, Dollman J, Dale MJ, Withers RT, Hill AM. Validation of DXA body composition estimates in obese men and women. *Obesity*. 2009;17(4):821-826.

25. Santos DA, Silva AM, Matias CN, Fields DA, Heymsfield SB, Sardinha LB. Accuracy of DXA in estimating body composition changes in elite athletes using a four compartment model as the reference method. *Nutr Metab (Lond)*. 2010;7:22.

26. Moon JR, Eckerson JM, Tobkin SE, et al. Estimating body fat in NCAA Division I female athletes: a five-compartment model validation of laboratory methods. *Eur J Appl Physiol*. 2009;105(1):119-130.

27. Forbes GB, Gallup J, Hursh JB. Estimation of total body fat from potassium-40 content. *Science*. 1961;133(3446):101-102.

28. Ellis KJ. Whole-body counting and neutron activation analysis. In: Heymsfield SB, Lohman TG, Wang Z, Going SB, eds. *Human Body Composition*. 2nd ed. Champaign, IL: Human Kinetics; 2005:51-62.

29. Cordain L, Johnson JE, Bainbridge CN, Wicker RE, Stockler JM. Potassium content of the fat free body in children. *J Sports Med Phys Fitness*. 1989;29(2):170-176.

30. Lohman TG. Applicability of body composition techniques and constants for children and youths. *Exerc Sport Sci Rev*. 1986;14:325-357.

31. Lohman TG, Pollock ML, Slaughter MH, Brandon LJ, Boileau RA. Methodological factors and the prediction of body fat in female athletes. *Med Sci Sports Exerc*. 1984;16(1):92-96.

32. Jackson AS, Pollock ML, Ward A. Generalized equations for predicting body density of women. *Med Sci Sports Exerc*. 1980;12(3):175-181.

33. Lohman TG. *Advances in Human Body Composition*. Champaign, IL: Human Kinetics; 1992.

34. Stevens J, Ou FS, Cai J, Heymsfield SB, Truesdale KP. Prediction of percent body fat measurements in Americans 8 years and older. *Int J Obes (Lond)*. 2016;40(4):587-594.

35. Pollock ML, Laughridge EE, Coleman B, Linnerud AC, Jackson A. Prediction of body density in young and middle-aged women. *J Appl Physiol*. 1975;38(4):745-749.

36. Pollock ML, Hickman T, Kendrick Z, Jackson A, Linnerud AC, Dawson G. Prediction of body density in young and middle-aged men. *J Appl Physiol*. 1976;40(3):300-304.

37. Jackson AS, Pollock ML. Practical assessment of body composition. *Phys Sportsmed.* 1985;13:76-90.

38. Peterson MJ, Czerwinski SA, Siervogel RM. Development and validation of skinfold-thickness prediction equations with a 4-compartment model. *Am J Clin Nutr.* 2003;77(5):1186-1191.

39. Lohman TG, Harris M, Teixeira PJ, Weiss L. Assessing body composition and changes in body composition. Another look at dual-energy X-ray absorptiometry. *Ann N Y Acad Sci.* 2000;904:45-54.

40. Hopkins WG. Bias in Bland-Altman but not regression validity analyses. *Sports Sci.* 2004;8:42-46.

41. Jackson AS, Ellis KJ, McFarlin BK, Sailors MH, Bray MS. Cross-validation of generalised body composition equations with diverse young men and women: the Training Intervention and Genetics of Exercise Response (TIGER) Study. *Br J Nutr.* 2009;101(6):871-878.

42. Schoeller DA, Tylavsky FA, Baer DJ, et al. QDR 4500A dual-energy X-ray absorptiometer underestimates fat mass in comparison with criterion methods in adults. *Am J Clin Nutr.* 2005;81(5):1018-1025.

43. Wong WW, Stuff JE, Butte NF, Smith EO, Ellis KJ. Estimating body fat in African American and white adolescent girls: a comparison of skinfold-thickness equations with a 4-compartment criterion model. *Am J Clin Nutr.* 2000;72(2):348-354.

44. Slaughter MH, Lohman TG, Boileau RA, et al. Skinfold equations for estimation of body fatness in children and youth. *Hum Biol.* 1988;60(5):709-723.

45. Stevens J, Cai J, Truesdale KP, Cuttler L, Robinson TN, Roberts AL. Percent body fat prediction equations for 8- to 17-year-old American children. *Pediatr Obes.* 2014;9(4):260-271.

46. Evans EM, Rowe DA, Misic MM, Prior BM, Arngrimsson SA. Skinfold prediction equation for athletes developed using a four-component model. *Med Sci Sports Exerc.* 2005;37(11):2006-2011.

47. Williams DP, Going SB, Lohman TG, Hewitt MJ, Haber AE. Estimation of body fat from skinfold thicknesses in middle-aged and older men and women: a multiple component approach. *Am J Hum Biol.* 1992;4:595-605.

48. Müller W, Horn M, Furhapter-Rieger A, et al. Body composition in sport: interobserver reliability of a novel ultrasound measure of subcutaneous fat tissue. *Br J Sports Med.* 2013;47(16):1036-1043.

49. Müller W, Horn M, Furhapter-Rieger A, et al. Body composition in sport: a comparison of a novel ultrasound imaging technique to measure subcutaneous fat tissue compared with skinfold measurement. *Br J Sports Med.* 2013;47(16):1028-1035.

50. Storchle P, Müller W, Sengeis M, et al. Standardized ultrasound measurement of subcutaneous fat patterning: high reliability and accuracy in groups ranging from lean to obese. *Ultrasound Med Biol.* 2017;43(2):427-438.

51. Müller W, Lohman TG, Stewart AD, et al. Subcutaneous fat patterning in athletes: selection of appropriate sites and standardisation of a novel ultrasound measurement technique: ad hoc working group on body composition, health and performance, under the auspices of the IOC Medical Commission. *Br J Sports Med.* 2016;50(1):45-54.

52. Heyward VH, Wagner DR. *Applied Body Composition Assessment.* 2nd ed. Champaign, IL: Human Kinetics; 2004.

53. Mialich MS, Sicchieri JMF, Junior AAJ. Analysis of body composition: a critical review of the use of bioelectrical impedance analysis. *Int J Clin Nutr.* 2014;2(1):1-10.

54. Bartok C, Schoeller DA, Randall Clark R, Sullivan JC, Landry GL. The effect of dehydration on wrestling minimum weight assessment. *Med Sci Sports Exerc.* 2004;36(1):160-167.

55. Clark RR, Bartok C, Sullivan JC, Schoeller DA. Is leg-to-leg BIA valid for predicting minimum weight in wrestlers? *Med Sci Sports Exerc.* 2005;37(6):1061-1068.

56. Jebb SA, Cole TJ, Doman D, Murgatroyd PR,

Prentice AM. Evaluation of the novel Tanita body-fat analyser to measure body composition by comparison with a four-compartment model. *Br J Nutr.* 2000;83(2):115-122.

57. Nunez C, Gallagher D, Visser M, Pi-Sunyer FX, Wang Z, Heymsfield SB. Bioimpedance analysis: evaluation of leg-to-leg system based on pressure contact footpad electrodes. *Med Sci Sports Exerc.* 1997;29(4):524-531.

58. Bera TK. Bioelectrical impedance methods for noninvasive health monitoring: a review. *J Med Eng.* 2014(381251Epub 2014 Jun 17).

59. Walter-Kroker A, Kroker A, Mattiucci-Guehlke M, Glaab T. A practical guide to bioelectrical impedance analysis using the example of chronic obstructive pulmonary disease. *Nutr J.* 2011;10:35.

60. Moon JR. Body composition in athletes and sports nutrition: an examination of the bioimpedance analysis technique. *Eur J Clin Nutr.* 2013;67(Suppl 1):S54-S59.

61. Elia M. Body composition by whole-body bioelectrical impedance and prediction of clinically relevant outcomes: overvalued or underused? *Eur J Clin Nutr.* 2013;67(Suppl 1):S60-S70.

62. Kushner RF, Schoeller DA. Estimation of total body water by bioelectrical impedance analysis. *Am J Clin Nutr.* 1986;44:417-424.

63. Chumlea WC, Guo SS, Kuczmarski RJ, et al. Body composition estimates from NHANES III bioelectrical impedance data. *Int J Obes Relat Metab Disord.* 2002;26(12):1596-1609.

64. Montagnese C, Williams JE, Haroun D, Siervo M, Fewtrell MS, Wells JC. Is a single bioelectrical impedance equation valid for children of wide ranges of age, pubertal status and nutritional status? Evidence from the 4-component model. *Eur J Clin Nutr.* 2013;67(Suppl 1):S34-39.

65. Kushner RF, Schoeller DA, Fjeld CR, Danford L. Is the impedance index (ht^2/R) significant in predicting total body water? *Am J Clin Nutr.* 1992;56:835-839.

66. Houtkooper LB, Going SB, Lohman TG, Roche AF, Van Loan M. Bioelectrical impedance estimation of fat-free body mass in children and youth: a cross-validation study. *J Appl Physiol.* 1992;72(1):366-373.

67. Wells JCK, Fuller NJ, Dewit O, Fewtrell MS, Elia M, Cole TJ. Four-component model of body composition in children: density and hydration of fat-free mass and comparison with simpler models. *Am J Clin Nutr.* 1999;69:904-912.

68. Bosy-Westphal A, Schautz B, Later W, Kehayias J, Gallagher D, Müller MJ. What makes a BIA equation unique? Validity of eight-electrode multifrequency BIA to estimate body composition in a healthy adult population. *Eur J Clin Nutr.* 2013;67(Suppl 1):S14-21.

69. Gallagher D, Visser M, Sepulveda D, Pierson RN, Harris T, Heymsfield SB. How useful is body mass index for comparison of body fatness across age, sex, and ethnic groups? *Am J Epidemiol.* 1996;143(3):228-239.

70. Rush EC, Freitas I, Plank LD. Body size, body composition and fat distribution: comparative analysis of European, Maori, Pacific Island and Asian Indian adults. *Br J Nutr.* 2009;102(4):632-641.

71. Deurenberg P, Deurenberg-Yap M, Guricci S. Asians are different from Caucasians and from each other in their body mass index/body fat per cent relationship. *Obes Rev.* 2002;3(3):141-146.

72. Fernandez JR, Heo M, Heymsfield SB, et al. Is percentage body fat differentially related to body mass index in Hispanic Americans, African Americans, and European Americans? *Am J Clin Nutr.* 2003;77(1):71-75.

73. Okorodudu DO, Jumean MF, Montori VM, et al. Diagnostic performance of body mass index to identify obesity as defined by body adiposity: a systematic review and meta-analysis. *Int J Obes.* 2010;34(5):791-799.

74. Gallagher D, Visser M, Sepulveda D, Pierson RN, Harris T, Heymsfield SB. How useful is body mass index for comparison of body fatness across age, sex, and ethnic groups? *Am J Epidemiol.* 1996;143(3):228-239.

75. Gallagher D, Ruts E, Visser M, et al. Weight stability masks sarcopenia in elderly men

and women. *Am J Physiol Endocrinol Metab.* 2000;279(2):E366-E375.

76. Javed A, Jumean M, Murad MH, et al. Diagnostic performance of body mass index to identify obesity as defined by body adiposity in children and adolescents: a systematic review and meta-analysis. *Pediatr Obes.* 2014.

77. Laurson KR, Eisenmann JC, Welk GJ. Development of youth percent body fat standards using receiver operating characteristic curves. *Am J Prev Med.* 2011;41(4 Suppl 2):S93-S99.

78. Laurson KR, Eisenmann JC, Welk GJ. Body fat percentile curves for U.S. children and adolescents. *Am J Prev Med.* 2011;41(4 Suppl 2):S87-S92.

79. Laurson KR, Eisenmann JC, Welk GJ. Body mass index standards based on agreement with health-related body fat. *Am J Prev Med.* 2011;41(4 Suppl 2):S100-S105.

80. Going SB, Lohman TG, Cussler EC, Williams DP, Morrison JA, Horn PS. Percent body fat and chronic disease risk factors in U.S. children and youth. *Am J Prev Med.* 2011;41(4 Suppl 2):S77-S86.

81. Sundgot-Borgen J, Meyer NL, Lohman TG, et al. How to minimise the health risks to athletes who compete in weight-sensitive sports review and position statement on behalf of the Ad Hoc Research Working Group on Body Composition, Health and Performance, under the auspices of the IOC Medical Commission. *Br J Sports Med.* 2013;47(16):1012-1022.

Chapter 8

1. Stewart AD, Sutton L. *Body Composition in Sport, Exercise and Health.* Abingdon, UK: Routledge; 2012.

2. Lohman TG. *Advances in Body Composition Assessment.* Champaign, IL: Human Kinetics; 1992.

3. Foreyt JP, Poston WS. Consensus view on the role of dietary fat and obesity. *Am J Med.* 2002;113(Suppl 9B):S60-S62.

4. Carr DB, Utzschneider KM, Hull RL, et al. Intra-abdominal fat is a major determinant of the National Cholesterol Education Program Adult Treatment Panel III criteria for the metabolic syndrome. *Diabetes.* 2004;53(8):2087-2094.

5. Bacon L, Stern JS, Van Loan MD, Keim NL. Size acceptance and intuitive eating improve health for obese, female chronic dieters. *J Am Diet Assoc.* 2005;105(6):929-936.

6. Karelis AD, St-Pierre DH, Conus F, Rabasa-Lhoret R, Poehlman ET. Metabolic and body composition factors in subgroups of obesity: what do we know? *J Clin Endocrinol Metab.* 2004;89(6):2569-2575.

7. Störchle P, Müller W, Sengeis M, et al. Standardized ultrasound measurement of subcutaneous fat patterning: high reliability and accuracy in groups ranging from lean to obese. *Ultrasound Med Biol.* 2017;43(2):427-438.

8. Brown LD. Endocrine regulation of fetal skeletal muscle growth: impact on future metabolic health. *J Endocrinol.* 2014;221(2):R13-R29.

9. Kulkarni B, Hills AP, Byrne NM. Nutritional influences over the life course on lean body mass of individuals in developing countries. *Nutr Rev.* 2014;72(3):190-204.

10. Goodell LS, Wakefield DB, Ferris AM. Rapid weight gain during the first year of life predicts obesity in 2-3 year olds from a low-income, minority population. *J Community Health.* 2009;34(5):370-375.

11. Taveras EM, Rifas-Shiman SL, Belfort MB, Kleinman KP, Oken E, Gillman MW. Weight status in the first 6 months of life and obesity at 3 years of age. *Pediatrics.* 2009;123(4):1177-1183.

12. Monteiro PO, Victora CG. Rapid growth in infancy and childhood and obesity in later life: a systematic review. *Obes Rev.* 2005;6(2):143-154.

13. Yang Z, Huffman SL. Nutrition in pregnancy and early childhood and associations with obesity in developing countries. *Matern Child Nutr.* 2013;9(Suppl 1):105-119.

14. Lohman TG, Chen Z. Dual-energy X-ray absorptiometry. In: Heymsfield SB, Lohman TG, Wang ZM, Going SB, eds. *Human Body*

Composition. 2nd ed. Champaign, IL: Human Kinetics; 2005:63-77.

15. Williams MH. *Nutrition for Health, Fitness and Sport.* New York, NY: McGraw-Hill; 2007.

16. Cordain L, Eaton SB, Sebastian A, et al. Origins and evolution of the Western diet: health implications for the 21st century. *Am J Clin Nutr.* 2005;81(2):341-354.

17. Bauer J, Biolo G, Cederholm T, et al. Evidence-based recommendations for optimal dietary protein intake in older people: a position paper from the PROT-AGE Study Group. *J Am Med Dir Assoc.* 2013;14(8):542-559.

18. Lonsdale D. Crime and violence: a hypothetical explanation of its relationship with high calorie malnutrition. *J Advan Med.* 1994;7(3):171-180.

19. Wortsman J, Matsuoka LY, Chen TC, Lu Z, Holick MF. Decreased bioavailability of vitamin D in obesity. *Am J Clin Nutr.* 2000;72(3):690-693.

20. Williams DP, Going SB, Lohman TG, et al. Body fatness and risk for elevated blood pressure, total cholesterol, and serum lipoprotein ratios in children and adolescents. *Am J Public Health.* 1992;82(3):358-363.

21. Lohman T, Hingle M, Going SB. Assessment of body composition in children in 1989 (25 years ago). *Pediatr Exerc Sci.* 2013;25(4):573-590.

22. Freedman DS, Sherry B. The validity of BMI as an indicator of body fatness and risk among children. *Pediatrics.* 2009;124(Suppl 1):S23-S34.

23. Freedman DS, Ogden CL, Kit BK. Interrelationships between BMI, skinfold thicknesses, percent body fat, and cardiovascular disease risk factors among U.S. children and adolescents. *BMC Pediatr.* 2015;15:188.

24. Going SB, Lohman TG, Cussler EC, Williams DP, Morrison JA, Horn PS. Percent body fat and chronic disease risk factors in U.S. children and youth. *Am J Prev Med.* 2011;41(Suppl 2):S77-S86.

25. Sardinha LB, Teixeira PJ. Measuring adiposity and fat distribution in relation to health. In: Heymsfield SB, Lohman TG, Wang ZM, Going SB, eds. *Human Body Composition.* 2nd ed. Champaign, IL: Human Kinetics; 2005:177-202.

26. Malina RM. Variation in body composition associated with sex and ethnicity. In: Heymsfield SB, Lohman TG, Wang ZM, Going SB, eds. *Human Body Composition.* 2nd ed. Champaign, IL: Human Kinetics; 2005:271-298.

27. Lukaski HC. Assessing muscle mass. In: Heymsfield SB, Lohman TG, Wang ZM, Going SB, eds. *Human Body Composition.* 2nd ed. Champaign, IL: Human Kinetics; 2005:203-218.

28. Farr JN, Khosla S. Skeletal changes through the lifespan: from growth to senescence. *Nat Rev Endocrinol.* 2015;11(9):513-521.

29. Webber LS, Catellier DJ, Lytle LA, et al. Promoting physical activity in middle school girls: Trial of Activity for Adolescent Girls. *Am J Prev Med.* 2008;34(3):173-184.

30. Baird J, Fisher D, Lucas P, Kleijnen J, Roberts H, Law C. Being big or growing fast: systematic review of size and growth in infancy and later obesity. *BMJ.* 2005;331(7522):929.

31. Schmelzle HR, Fusch C. Body fat in neonates and young infants: validation of skinfold thickness versus dual-energy X-ray absorptiometry. *Am J Clin Nutr.* 2002;76(5):1096-1100.

32. Deierlein AL, Thornton J, Hull H, Paley C, Gallagher D. An anthropometric model to estimate neonatal fat mass using air displacement plethysmography. *Nutr Metab (Lond).* 2012;9:21.

33. Guo SS, Chumlea WC. Tracking of body mass index in children in relation to overweight in adulthood. *Am J Clin Nutr.* 1999;70(1):145S-148S.

34. Wolfe RR, Miller SL, Miller KB. Optimal protein intake in the elderly. *Clin Nutr.* 2008;27(5):675-684.

35. Evans WJ, Morley JE, Argiles J, et al. Cachexia: a new definition. *Clin Nutr.* 2008;27(6):793-799.

36. Heymsfield SB, McManus C, Smith J, Stevens V, Nixon DW. Anthropometric measurement of muscle mass: revised equations for calculating bone-free arm muscle area. *Am J Clin Nutr.* 1982;36:680-690.

37. Barac-Nieto M, Spurr GB, Lotero H, Maksud MG. Body composition in chronic undernutrition. *Am J Clin Nutr.* 1978;31(1):23-40.

38. Meyer NL, Sundgot-Borgen J, Lohman TG, et al. Body composition for health and performance: a survey of body composition assessment practice carried out by the Ad Hoc Research Working Group on Body Composition, Health and Performance under the auspices of the IOC Medical Commission. *Br J Sports Med.* 2013;47(16):1044-1053.

39. Obesity and overweight. World Health Organization website. www.who.int/mediacentre/factsheets/fs311/en/. Accessed February 10, 2016.

40. Mountjoy M, Sundgot-Borgen J, Burke L, et al. The IOC consensus statement: beyond the Female Athlete Triad—Relative Energy Deficiency in Sport (RED-S). *Br J Sports Med.* 2014;48(7):491-497.

41. Sundgot-Borgen J, Meyer NL, Lohman TG, et al. How to minimise risks for weight sensitive sports: review and position statement on behalf of the Ad Hoc Research Working Group on Body Composition, Health and Performance, under the auspices of the IOC Medical Commission. *Br J Sports Med.* 2013;47:1012-1022.

42. Ackland T, Lohman T, Sundgot-Borgen J, et al. Current status of body composition assessment in sport: review and position statement on behalf of the Ad Hoc Research Working Group on Body Composition Health and Performance, under the auspices of the I.O.C. Medical Commission. *Sports Med.* 2012;42(3):227-249.

43. Müller W, Lohman TG, Stewart AD, et al. Subcutaneous fat patterning in athletes: selection of appropriate sites and standardisation of a novel ultrasound measurement technique: ad hoc working group on body composition, health and performance, under the auspices of the IOC Medical Commission. *Br J Sports Med.* 2016;50(1):45-54.

44. Garthe I, Raastad T, Refsnes PE, Koivisto A, Sundgot-Borgen J. Effect of two different weight-loss rates on body composition and strength and power-related performance in elite athletes. *Int J Sport Nutr Exerc Metab.* 2011;21(2):97-104.

45. Sundgot-Borgen J, Garthe I. Elite athletes in aesthetic and Olympic weight-class sports and the challenge of body weight and body compositions. *J Sports Sci.* 2011;29(Suppl 1):S101-S114.

46. Mettler S, Mitchell N, Tipton KD. Increased protein intake reduces lean body mass loss during weight loss in athletes. *Med Sci Sports Exerc.* 2010;42(2):326-337.

47. Helms ER, Zinn C, Rowlands DS, Brown SR. A systematic review of dietary protein during caloric restriction in resistance trained lean athletes: a case for higher intakes. *Int J Sport Nutr Exerc Metab.* 2014;24(2):127-138.

48. Heyward VH, Wagner DR. *Applied Body Composition Assessment.* 2nd ed. Champaign, IL: Human Kinetics; 2004.

49. Tipton KD, Wolfe RR. Exercise, protein metabolism, and muscle growth. *Int J Sport Nutr Exerc Metab.* 2001;11(1):109-132.

50. Phillips SM, Tipton KD, Ferrando AA, Wolfe RR. Resistance training reduces the acute exercise-induced increase in muscle protein turnover. *Am J Physiol.* 1999;276(1 Pt 1):E118-E124.

51. Lohman T, Going S, Pamenter R, et al. Effects of resistance training on regional and total bone mineral density in premenopausal women: a randomized prospective study. *J Bone Miner Res.* 1995;10(7):1015-1024.

52. Lohman T. Exercise and bone mineral density. *Quest.* 1995;47:354-361.

53. Heymsfield SB, Lohman TG, Wang Z, Going SB, eds. *Human Body Composition.* 2nd ed. Champaign, IL: Human Kinetics; 2005.

54. American Psychiatric Association. *Diagnostic and Statistical Manual for Mental Disorders: DSM-IV.* 4th ed. Washington, DC: American Psychiatric Association; 1994.

55. Nattiv A, Loucks AB, Manore MM, et al. American College of Sports Medicine position stand: the female athlete triad. *Med Sci Sports Exerc.* 2007;39(10):1867-1882.

56. Torstveit MK, Aagedal-Mortensen K, Stea TH. More than half of high school students report disordered eating: a cross sectional study among Norwegian boys and girls. *PLoS One.* 2015;10(3):e0122681.

57. Smink FR, van Hoeken D, Oldehinkel AJ, Hoek HW. Prevalence and severity of DSM-5 eating disorders in a community cohort of adolescents. *Int J Eat Disord.* 2014;47(6):610-619.

58. Neumark-Sztainer D, Wall M, Larson NI, Eisenberg ME, Loth K. Dieting and disordered eating behaviors from adolescence to young adulthood: findings from a 10-year longitudinal study. *J Am Diet Assoc.* 2011;111(7):1004-1011.

59. Sundgot-Borgen J, Torstveit MK. Prevalence of eating disorders in elite athletes is higher than in the general population. *Clin J Sport Med.* 2004;14(1):25-32.

60. American Dietetic Association. Practice paper of the American Dietetic Association: nutrition intervention in the treatment of eating disorders. *J Am Diet Assoc.* 2011;111:1236-1241.

61. American Psychiatric Association. *American Psychiatric Association Practice Guidelines for the Treatment of Psychiatric Disorders: Compendium 2006.* Arlington, VA: American Psychiatric Publishing; 2006.

62. Lund BC, Hernandez ER, Yates WR, Mitchell JR, McKee PA, Johnson CL. Rate of inpatient weight restoration predicts outcome in anorexia nervosa. *Int J Eat Disord.* 2009;42(4):301-305.

63. Carter JC, Mercer-Lynn KB, Norwood SJ, et al. A prospective study of predictors of relapse in anorexia nervosa: implications for relapse prevention. *Psychiatry Res.* 2012;200(2-3):518-523.

64. Probst M, Goris M, Vandereycken W, Van Coppenolle H. Body composition of anorexia nervosa patients assessed by underwater weighing and skinfold-thickness measurements before and after weight gain. *Am J Clin Nutr.* 2001;73(2):190-197.

65. Salisbury JJ, Levine AS, Crow SJ, Mitchell JE. Refeeding, metabolic rate, and weight gain in anorexia nervosa: a review. *Int J Eat Disord.* 1995;17(4):337-345.

66. Scalfi L, Polito A, Bianchi L, et al. Body composition changes in patients with anorexia nervosa after complete weight recovery. *Eur J Clin Nutr.* 2002;56(1):15-20.

67. Fernandez-del-Valle M, Larumbe-Zabala E, Villasenor-Montarroso A, et al. Resistance training enhances muscular performance in patients with anorexia nervosa: a randomized controlled trial. *Int J Eat Disord.* 2014;47(6):601-609.

68. Fernandez-del-Valle M, Larumbe-Zabala E, Graell-Berna M, Perez-Ruiz M. Anthropometric changes in adolescents with anorexia nervosa in response to resistance training. *Eat Weight Disord.* 2015;20(3):311-317.

69. Bratland-Sanda S, Martinsen EW, Sundgot-Borgen J. Changes in physical fitness, bone mineral density and body composition during inpatient treatment of underweight and normal weight females with longstanding eating disorders. *Int J Environ Res Public Health.* 2012;9(1):315-330.

70. Zuckerman-Levin N, Hochberg Z, Latzer Y. Bone health in eating disorders. *Obes Rev.* 2014;15(3):215-223.

71. Modan-Moses D, Levy-Shraga Y, Pinhas-Hamiel O, et al. High prevalence of vitamin D deficiency and insufficiency in adolescent inpatients diagnosed with eating disorders. *Int J Eat Disord.* 2015;48(6):607-614.

72. Fogelholm M, Sievanen H, Heinonen A, et al. Association between weight cycling history and bone mineral density in premenopausal women. *Osteoporos Int.* 1997;7(4):354-358.

73. Shuster A, Patlas M, Pinthus JH, Mourtzakis M. The clinical importance of visceral adiposity: a critical review of methods for visceral adipose tissue analysis. *Br J Radiol.* 2012;85(1009):1-10.

74. Kaul S, Rothney MP, Peters DM, et al. Dual-energy X-ray absorptiometry for quantification of visceral fat. *Obesity (Silver Spring).* 2012;20(6):1313-1318.

75. De Lucia Rolfe E, Sleigh A, Finucane FM, et al. Ultrasound measurements of visceral and

subcutaneous abdominal thickness to predict abdominal adiposity among older men and women. *Obesity (Silver Spring)*. 2010;18(3):625-631.

76. Philipsen A, Jørgensen ME, Vistisen D, et al. Associations between ultrasound measures of abdominal fat distribution and indices of glucose metabolism in a population at high risk of type 2 diabetes: the ADDITION-PRO study. *PLoS One*. 2015;10(4):e0123062.

77. Wagner DR. Ultrasound as a tool to assess body fat. http://dx.doi.org/10.1155/2013/280713. *J Obes*. 2013;2013:1-9.

78. Lee DH, Park KS, Ahn S, et al. Comparison of abdominal visceral adipose tissue area measured by computed tomography with that estimated by bioelectrical impedance analysis method in Korean subjects. *Nutrients*. 2015;7(12):10513-10524.

79. Tomiyama AJ, Hunger JM, Nguyen-Cuu J, Wells C. Misclassification of cardiometabolic health when using body mass index categories in NHANES 2005-2012. *Int J Obes (Lond)*. 2016;40(5):883-886.

80. ACSM. *ACSM's Guidelines for Exercise Testing and Prescription*. 10th ed. China: Wolters Kluwer; 2017.

81. Despres JP. Abdominal obesity and cardiovascular disease: is inflammation the missing link? *Can J Cardiol*. 2012;28(6):642-652.

82. Harris MM. Obesity and fat distribution. In: Caballero B, Allen L, Prentice AM, eds. *Encyclopedia of Human Nutrition*. 1st ed. Kidlington, UK: Academic Press; 1998:1973.

83. Pinho CPS, Diniz ADS, de Arruda IKG, Leite A, Petribu MMV, Rodrigues IG. Predictive models for estimating visceral fat: the contribution from anthropometric parameters. *PLoS One*. 2017;12(7):e0178958.

84. Swainson MG, Batterham AM, Tsakirides C, Rutherford ZH, Hind K. Prediction of whole-body fat percentage and visceral adipose tissue mass from five anthropometric variables. *PLoS One*. 2017;12(5):e0177175.

85. Egger G, Dobson A. Clinical measures of obesity and weight loss in men. *Int J Obes Relat Metab Disord*. 2000;24(3):354-357.

86. Pourhassan M, Schautz B, Braun W, Gluer CC, Bosy-Westphal A, Müller MJ. Impact of body-composition methodology on the composition of weight loss and weight gain. *Eur J Clin Nutr*. 2013;67(5):446-454.

87. Heymsfield SB, Gonzalez MC, Shen W, Redman L, Thomas D. Weight loss composition is one-fourth fat-free mass: a critical review and critique of this widely cited rule. *Obes Rev*. 2014;15(4):310-321.

88. Evans EM, Mojtahedi MC, Thorpe MP, Valentine RJ, Kris-Etherton PM, Layman DK. Effects of protein intake and gender on body composition changes: a randomized clinical weight loss trial. *Nutr Metab (Lond)*. 2012;9(1):55.

89. Beavers KM, Ambrosius WT, Rejeski WJ, et al. Effect of exercise type during intentional weight loss on body composition in older adults with obesity. *Obesity (Silver Spring)*. 2017;25(11):1823-1829.

90. Dulloo AG, Jacquet J, Montani JP, Schutz Y. How dieting makes the lean fatter: from a perspective of body composition autoregulation through adipostats and proteinstats awaiting discovery. *Obes Rev*. 2015;16(Suppl 1):25-35.

91. Backx EM, Tieland M, Borgonjen-van den Berg KJ, Claessen PR, van Loon LJ, de Groot LC. Protein intake and lean body mass preservation during energy intake restriction in overweight older adults. *Int J Obes (Lond)*. 2016;40(2):299-304.

92. Goss AM, Goree LL, Ellis AC, et al. Effects of diet macronutrient composition on body composition and fat distribution during weight maintenance and weight loss. *Obesity (Silver Spring)*. 2013;21(6):1139-1142.

93. Liebman M. When and why carbohydrate restriction can be a viable option. *Nutrition*. 2014;30(7-8):748-754.

94. Shapses SA, Von Thun NL, Heymsfield SB, et al. Bone turnover and density in obese premeno-

pausal women during moderate weight loss and calcium supplementation. *J Bone Miner Res.* 2001;16(7):1329-1336.

95. Bowen J, Noakes M, Clifton PM. A high dairy protein, high-calcium diet minimizes bone turnover in overweight adults during weight loss. *J Nutr.* 2004;134(3):568-573.

96. Rector RS, Loethen J, Ruebel M, Thomas TR, Hinton PS. Serum markers of bone turnover are increased by modest weight loss with or without weight-bearing exercise in overweight premenopausal women. *Appl Physiol Nutr Metab.* 2009;34(5):933-941.

97. Labouesse MA, Gertz ER, Piccolo BD, et al. Associations among endocrine, inflammatory, and bone markers, body composition and physical activity to weight loss induced bone loss. *Bone.* 2014;64:138-146.

98. Ding J, Kritchevsky SB, Newman AB, et al. Effects of birth cohort and age on body composition in a sample of community-based elderly. *Am J Clin Nutr.* 2007;85(2):405-410.

99. Cesari M, Kritchevsky SB, Baumgartner RN, et al. Sarcopenia, obesity, and inflammation: results from the Trial of Angiotensin Converting Enzyme Inhibition and Novel Cardiovascular Risk Factors study. *Am J Clin Nutr.* 2005;82(2):428-434.

100. Visser M, Langlois J, Guralnik JM, et al. High body fatness, but not low fat-free mass, predicts disability in older men and women: the Cardiovascular Health Study. *Am J Clin Nutr.* 1998;68(3):584-590.

101. Kuk JL, Saunders TJ, Davidson LE, Ross R. Age-related changes in total and regional fat distribution. *Ageing Res Rev.* 2009;8(4):339-348.

102. Franklin RM, Ploutz-Snyder L, Kanaley JA. Longitudinal changes in abdominal fat distribution with menopause. *Metabolism.* 2009;58(3):311-315.

103. Baumgartner RN, Rhyne RL, Garry PJ, Heymsfield SB. Imaging techniques and anatomical body composition in aging. *J Nutr.* 1993;123(2 Suppl):444-448.

104. Poehlman ET, Toth MJ, Bunyard LB, et al. Physiological predictors of increasing total and central adiposity in aging men and women. *Arch Intern Med.* 1995;155(22):2443-2448.

105. Gallagher D, Visser M, De Meersman RE, et al. Appendicular skeletal muscle mass: effects of age, gender, and ethnicity. *J Appl Physiol.* 1997;83(1):229-239.

106. Frontera WR, Hughes VA, Fielding RA, Fiatarone MA, Evans WJ, Roubenoff R. Aging of skeletal muscle: a 12-yr longitudinal study. *J Appl Physiol.* 2000;88(4):1321-1326.

107. Deschenes MR. Effects of aging on muscle fibre type and size. *Sports Med.* 2004;34(12):809-824.

108. Batsis JA, Mackenzie TA, Jones JD, Lopez-Jimenez F, Bartels SJ. Sarcopenia, sarcopenic obesity and inflammation: results from the 1999-2004 National Health and Nutrition Examination Survey. *Clin Nutr.* 2016.

109. Visser M, Pahor M, Taaffe DR, et al. Relationship of interleukin-6 and tumor necrosis factor-alpha with muscle mass and muscle strength in elderly men and women: the Health ABC Study. *J Gerontol A Biol Sci Med Sci.* 2002;57(5):M326-M332.

110. Baumgartner RN, Wayne SJ, Waters DL, Janssen I, Gallagher D, Morley JE. Sarcopenic obesity predicts instrumental activities of daily living disability in the elderly. *Obes Res.* 2004;12(12):1995-2004.

111. Visser M, Goodpaster BH, Kritchevsky SB, et al. Muscle mass, muscle strength, and muscle fat infiltration as predictors of incident mobility limitations in well-functioning older persons. *J Gerontol A Biol Sci Med Sci.* 2005;60(3):324-333.

112. Goodpaster BH, Kelley DE. Obesity and diabetes: body composition determinants of insulin resistance. In: Heymsfield S, Lohman T, Wang Z, Going S, eds. *Human Body Composition.* 2nd ed. Champaign, IL: Human Kinetics; 2005:365-375.

113. Blunt BA, Klauber MR, Barrett-Connor EL, Edelstein SL. Sex differences in bone mineral density in 1653 men and women in the sixth

through tenth decades of life: the Rancho Bernardo Study. *J Bone Miner Res.* 1994;9(9):1333-1338.

114. Lloyd JT, Alley DE, Hochberg MC, et al. Changes in bone mineral density over time by body mass index in the Health ABC study. *Osteoporos Int.* 2016.

115. Farhat GN, Newman AB, Sutton-Tyrrell K, et al. The association of bone mineral density measures with incident cardiovascular disease in older adults. *Osteoporos Int.* 2007;18(7):999-1008.

116. World Health Organization. *Assessment of fracture risk and its application to screening for postmenopausal osteoporosis: report of a WHO Study Group.* Geneva; 1994.

117. World Health Organization. *WHO scientific group on the assessment of osteoporosis at primary health care level: summary meeting report.* Brussels; 2007.

118. Behnke AR, Feen BG, Welham WC. The specific gravity of healthy men. Body weight divided by volume as an index of obesity. *JAMA.* 1942;118:495-498.

119. Pace N, Rathbun EN. Studies on body composition: body water and chemically combined nitrogen content in relation to fat content. *J Biol Chem.* 1945;158:685-691.

120. Visser M, Gallagher D, Deurenberg P, Wang J, Pierson RN Jr., Heymsfield SB. Density of fat-free body mass: relationship with race, age, and level of body fatness. *Am J Physiol.* 1997;272(5 Pt 1):E781-E787.

121. Baumgartner RN, Heymsfield SB, Lichtman S, Wang J, Pierson RN, Jr. Body composition in elderly people: effect of criterion estimates on predictive equations. *Am J Clin Nutr.* 1991;53(6):1345-1353.

122. Heymsfield SB, Wang J, Lichtman S, Kamen Y, Kehayias J, Pierson RN Jr. Body composition in elderly subjects: a critical appraisal of clinical methodology. *Am J Clin Nutr.* 1989;50(5 Suppl):1167-1175; discussion 1231-1165.

123. Yee AJ, Fuerst T, Salamone L, et al. Calibration and validation of an air-displacement plethys-mography method for estimating percentage body fat in an elderly population: a comparison among compartmental models. *Am J Clin Nutr.* 2001;74(5):637-642.

124. Withers RT, LaForgia J, Pillans RK, et al. Comparisons of two-, three-, and four-compartment models of body composition analysis in men and women. *J Appl Physiol.* 1998;85(1):238-245.

125. Pietrobelli A, Wang Z, Formica C, Heymsfield SB. Dual-energy X-ray absorptiometry: fat estimation errors due to variation in soft tissue hydration. *Am J Physiol.* 1998;274(5 Pt 1):E808-E816.

126. Toombs RJ, Ducher G, Shepherd JA, De Souza MJ. The impact of recent technological advances on the trueness and precision of DXA to assess body composition. *Obesity (Silver Spring).* 2012;20(1):30-39.

127. Visser M, Kritchevsky SB, Goodpaster BH, et al. Leg muscle mass and composition in relation to lower extremity performance in men and women aged 70 to 79: the health, aging and body composition study. *J Am Geriatr Soc.* 2002;50(5):897-904.

128. Goodpaster BH, Park SW, Harris TB, et al. The loss of skeletal muscle strength, mass, and quality in older adults: the Health, Aging and Body Composition Study. *J Gerontol A Biol Sci Med Sci.* 2006;61(10):1059-1064.

129. Newman AB, Lee JS, Visser M, et al. Weight change and the conservation of lean mass in old age: the Health, Aging and Body Composition Study. *Am J Clin Nutr.* 2005;82(4):872-878; quiz 915-876.

130. Cussler EC, Going SB, Houtkooper LB, et al. Exercise frequency and calcium intake predict four-year bone changes in postmenopausal women. *Osteoporos Int.* 2005;16(12):2129-2141.

131. Milliken LA, Cussler E, Zeller RA, et al. Changes in soft tissue composition are the primary predictors of 4-year bone mineral density changes in postmenopausal women. *Osteoporos Int.* 2009;20:347-354.

132. Heymsfield SB, Nunez C, Testolin C, Gallagher D. Anthropometry and methods of body

composition measurement for research and field application in the elderly. *Eur J Clin Nutr.* 2000;54(Suppl 3):S26-S32.

133. Lean ME, Han TS, Deurenberg P. Predicting body composition by densitometry from simple anthropometric measurements. *Am J Clin Nutr.* 1996;63(1):4-14.

134. Durnin JV, Womersley J. Body fat assessed from total body density and its estimation from skinfold thickness: measurements on 481 men and women aged from 16 to 72 years. *Br J Nutr.* 1974;32(1):77-97.

135. Kyle UG, Genton L, Hans D, Pichard C. Validation of a bioelectrical impedance analysis equation to predict appendicular skeletal muscle mass (ASMM). *Clin Nutr.* 2003;22(6):537-543.

136. Dos Santos L, Cyrino ES, Antunes M, Santos DA, Sardinha LB. Changes in phase angle and body composition induced by resistance training in older women. *Eur J Clin Nutr.* 2016;70(12):1408-1413.

137. Norman K, Stobaus N, Pirlich M, Bosy-Westphal A. Bioelectrical phase angle and impedance vector analysis: clinical relevance and applicability of impedance parameters. *Clin Nutr.* 2012;31(6):854-861.

138. Earthman CP. Body composition tools for assessment of adult malnutrition at the bedside: a tutorial on research considerations and clinical applications. *JPEN J Parenter Enteral Nutr.* 2015;39(7):787-822.

139. Plank LD, Li A. Bioimpedance illness marker compared to phase angle as a predictor of malnutrition in hospitalised patients [abstract]. *Clin Nutr.* 2013;32(Suppl 1):S85.

140. Lukaski HC, Kyle UG, Kondrup J. Assessment of adult malnutrition and prognosis with bioelectrical impedance analysis: phase angle and impedance ratio. *Curr Opin Clin Nutr Metab Care.* 2017;20(5):330-339.

141. Onofriescu M, Hogas S, Voroneanu L, et al. Bioimpedance-guided fluid management in maintenance hemodialysis: a pilot randomized controlled trial. *Am J Kidney Dis.* 2014;64(1):111-118.

142. Lemos T, Gallagher D. Current body composition measurement techniques. *Curr Opin Endocrinol Diabetes Obes.* 2017;24(5):310-314.

143. Toro-Ramos T, Paley C, Pi-Sunyer FX, Gallagher D. Body composition during fetal development and infancy through the age of 5 years. *Eur J Clin Nutr.* 2015;69(12):1279-1289.

144. Shepherd JA, Ng BK, Fan B, et al. Modeling the shape and composition of the human body using dual energy X-ray absorptiometry images. *PLoS One.* 2017;12(4):e0175857.

INDEX

Note: The italicized *f* and *t* following page numbers refer to figures and tables, respectively.

ABOUT THE EDITORS

The **American College of Sports Medicine (ACSM)**, founded in 1954, is the largest sports medicine and exercise science organization in the world. With more than 50,000 members and certified professionals worldwide, ACSM is dedicated to improving health through science, education, and medicine. ACSM members work in a range of medical specialties, allied health professions, and scientific disciplines. Members are committed to the diagnosis, treatment, and prevention of sport-related injuries and the advancement of the science of exercise. The ACSM promotes and integrates scientific research, education, and practical applications of sports medicine and exercise science to maintain and enhance physical performance, fitness, health, and quality of life.

Timothy G. Lohman, PhD, is a professor emeritus at the University of Arizona and is widely considered a leading scientist in the field of body composition assessment. His research includes serving as principal investigator (PI) of both the TAAG (Trial of Activity for Adolescent Girls) study—a collaborative multicenter study focused on physical activity of adolescent girls—and the Bone Estrogen Strength Training (BEST) study. He was co-PI of the Pathways Study, a collaborative study (by the National Heart, Lung, and Blood Institute; four field centers; and a coordinating center) designed to prevent obesity in Native American children. Lohman served as a consultant to the Women's Health Initiative (WHI) Vanguard Center and Health ABC study of long-term aging, and he was an advisor on youth fitness for the Cooper Institute and is a member of ACSM. He previously served as the director of the Center for Physical Activity and Nutrition at the University of Arizona.

Lohman's additional works, published by Human Kinetics, include his co-edited *Human Body Composition, Second Edition*; his authored monograph, "Advances in Body Composition Assessment"; and his co-edited *Anthropometric Standardization Reference Manual*. His research in body composition helped to establish the chemical immaturity of children using the multicomponent model.

His photo is courtesy of J'Fleur Lohman.

Laurie A. Milliken, PhD, FACSM, is an associate professor and former chair of the exercise and health sciences department at the University of Massachusetts Boston. In the New England chapter of the American College of Sports Medicine (NEACSM), she has served as a state representative, an executive committee member, the Continuing Education Committee chair, and president, and she has been an active member since 1998. Nationally, she has served on the ACSM Research Awards Committee and is also an editorial board member of *ACSM's Health & Fitness Journal*. She is currently a peer reviewer for leading scientific journals such as *Medicine & Science in Sports & Exercise*, the *Journal of Applied Physiology*, and the *European Journal of Applied Physiology*. She has been a member of ACSM since 1994 and has presented her research at many annual meetings. Her research interests include the regulation of body composition in response to exercise throughout the lifespan. She has received NIH funding for her work and is also a fellow of the American College of Sports Medicine.

Her photo is courtesy of John Maciel Photography.

CONTRIBUTORS

Jennifer W. Bea, PhD
University of Arizona Cancer Center
Tucson, Arizona

Robert M. Blew, MS
University of Arizona
Tucson, Arizona

Leslie Jerome Brandon, PhD, FACSM
Georgia State University
Atlanta, Georgia

Kirk Cureton, PhD, FACSM
University of Georgia
Athens, Georgia

Margaret Harris, PhD
University of Colorado
Colorado Springs, Colorado

Nuwanee Kirihennedige, MS, RD
University of Colorado
United States Olympic Committee
Colorado Springs, Colorado

Vinson Lee, MS
University of Arizona
Tucson, Arizona

Nanna Lucia Meyer, PhD, FACSM
University of Colorado
Colorado Springs, Colorado

Alba Reguant-Closa, MS, RD
University of Vic–Central University of Catalonia,
    Barcelona, Spain
University of Andorra, Sant Julià de Lòria, Andorra

Vanessa Risoul-Salas, MSc, RD
Hospital Ángeles del Pedregal
Mexico City, Mexico

Luis B. Sardinha, PhD
University of Lisbon
Lisbon, Portugal

REVIEWERS

Robert Berry, MS, ACSM-CEP, RCEP, EIM
Henry Ford Health System
Hebron, Connecticut

Gregory B. Dwyer, PhD, FACSM, ACSM-CEP, PD, ETT, EIM
East Stroudsburg University
East Stroudsburg, Pennsylvania

Michael Andres Figueroa, EdD
William Paterson University
Wayne, New Jersey

Elizabeth K. Lenz, PhD
The College at Brockport–SUNY, KSSPE
Brockport, Wisconsin

Thomas P. Mahady, MS
Hackensack University Medical Center
Hackensack, New Jersey

Emily J. Sauers, PhD
East Stroudsburg University
East Stroudsburg, Pennsylvania

Paul M. Sorace, MS, FACSM, ACSM-CEP, RCEP
Hackensack University Medical Center
Hackensack, New Jersey

Brianna Wells, MS
Sacred Heart University
Fairfield, Connecticut

You read the book—now complete an exam to earn continuing education credit.

Congratulations on successfully preparing for this continuing education exam!

If you would like to earn CE credit, please visit

www.HumanKinetics.com/CE-Exam-Access

for complete instructions on how to access your exam.
Take advantage of a discounted rate by entering
promo code **ABCA2019** when prompted.

HUMAN KINETICS